NATURE SCIENCE AND SUSTAINABLE TECHNOLOGY

NATURE SCIENCE AND SUSTAINABLE TECHNOLOGY

M. RAFIQUL ISLAM
EDITOR

Nova Science Publishers, Inc.
New York

Copyright © 2008 by Nova Science Publishers, Inc.

All rights reserved. No part of this book may be reproduced, stored in a retrieval system or transmitted in any form or by any means: electronic, electrostatic, magnetic, tape, mechanical photocopying, recording or otherwise without the written permission of the Publisher.

For permission to use material from this book please contact us:
Telephone 631-231-7269; Fax 631-231-8175
Web Site: http://www.novapublishers.com

NOTICE TO THE READER

The Publisher has taken reasonable care in the preparation of this book, but makes no expressed or implied warranty of any kind and assumes no responsibility for any errors or omissions. No liability is assumed for incidental or consequential damages in connection with or arising out of information contained in this book. The Publisher shall not be liable for any special, consequential, or exemplary damages resulting, in whole or in part, from the readers' use of, or reliance upon, this material.

Independent verification should be sought for any data, advice or recommendations contained in this book. In addition, no responsibility is assumed by the publisher for any injury and/or damage to persons or property arising from any methods, products, instructions, ideas or otherwise contained in this publication.

This publication is designed to provide accurate and authoritative information with regard to the subject matter covered herein. It is sold with the clear understanding that the Publisher is not engaged in rendering legal or any other professional services. If legal or any other expert assistance is required, the services of a competent person should be sought. FROM A DECLARATION OF PARTICIPANTS JOINTLY ADOPTED BY A COMMITTEE OF THE AMERICAN BAR ASSOCIATION AND A COMMITTEE OF PUBLISHERS.

LIBRARY OF CONGRESS CATALOGING-IN-PUBLICATION DATA

Nature science and sustainable technology research progress / M. Rafiqul Islam, ed.
  p. cm.
 ISBN 978-1-60456-310-8 (hardcover)
 1. Environmental protection. 2. Green technology. 3. Environmental sciences. 4. Human ecology. I. Islam, Rafiqul, 1959-
TD170.3.N38 2007a
628--dc22
                    2007048723

Published by Nova Science Publishers, Inc. New York

CONTENTS

Preface		vii
Chapter 1	Truth Is Knowledge, Knowledge is Freedom, and Freedom is Peace. So, what is the Problem? *M. R. Islam*	1
Chapter 2	Development and Application of Criteria for True Sustainability *M. Ibrahim Khan*	5
Chapter 3	The Engineering Approach Versus the Mathematical Approach in Developing Reservoir Simulators *J. H. Abou-Kassem*	39
Chapter 4	Surface Chemistry of Atlantic Cod Scale *A. Basu, R. L. White, M. D. Lumsden, P. Bishop, S. Butt, S. Mustafiz and M. R. Islam*	73
Chapter 5	Reversing Global Warming *A. B. Chhetri and M. R. Islam*	83
Chapter 6	The Adomian Decomposition Method on Solutions of Non-Linear Partial Differential Equations *S. H. Mousavizadegan, S. Mustafiz and M. Rahman*	119
Chapter 7	A Novel Sustainable Combined Heating/Cooling/Refrigeration System *M. M. Khan, D. Prior and M. R. Islam*	137
Index		167

Preface

Nature thrives on diversity and flexibility, gaining strength from heterogeneity, whereas the quest for homogeneity seems to motivate much of modern engineering. Nature is non-linear and inherently promotes multiplicity of solutions. This new book presents new and original research on true sustainability and technology development.

Chapter 1

TRUTH IS KNOWLEDGE, KNOWLEDGE IS FREEDOM, AND FREEDOM IS PEACE. SO, WHAT IS THE PROBLEM?

M. R. Islam
Civil and Resource Engineering Department,
Dalhousie University, Halifax, Canada

The Information Age is synonymous with an overflow, a superflux, of "information". Information is necessary for traveling the path of knowledge, leading to the truth. Truth sets one free. Freedom is peace. Yet here a horrific contradiction leaps out to grab one and all by the throat: of all the characteristics that can be said to characterize the Information Age, freedom is not one of them.

While the lack of freedom and lack of access to the truth is quite clear in social and political matters, conditions are in fact – contrary to what many assume – hardly better in science or technology development. The world has borne witness to the fact that in last 50 years there has been a 50-times increase per capita in cases of non-genetic diabetes, cancer, myopia, immune deficiency, asthma, and obesity. These stunning increases bear the most ominous link to modern lifestyle - namely, a 50-times increase in per-capita use of sugar ('refined' carbohydrate), plastic ('wrinkle-free' leather, 'durable' wood, 'cheap' water container), fertilizer ('refined' biomass), spirit ('refined' alcohol), cigarettes ('refined' tobacco), and numerous chemicals (ranging from 'preservatives' to 'anti-biotics' and 'pesticides').

However, few have ventured to find the reason behind what Robert Curl (Nobel Laureate in Chemistry) called 'our technological disaster', of which the items mentioned in the brief catalog above represent but symptoms. Talk of moral decline is all the rage, yet few saw 'the marvels of science' as the driver of the downward spiraling rush into ever more anti-Nature schemes. This journal stands out precisely as one of those rare efforts into exploring *why* technologies failed humanity and science betrayed our conscience.

The first paper explains why human intervention has potentially propelled us in the path of self-destruction, all in the name of … "sustainable development". This paper unravels the mysteries of sustainable development – the buzz-phrase that has become an instrument for creating yet another recipe for injustice of infinite magnitude. The paper clearly shows why previous criteria failed to predict the current human condition and why these criteria cannot be trusted to salvage us from the disaster that was created by the same criteria to begin with. Criticism of the current *modus operandi* is not rare – but it is also not sufficient. The paper finds explanations for every mechanism involved in developing technologies and scrutinizes the background theories, including fundamental ones that have been taken too long for granted. It lays the foundation for solutions that are innovative, environmentally appealing, and socially responsible. Such an approach is a giant step forward toward undoing some considerable portion of the damage already inflicted on our planet, setting the stage for further papers that will supplement this notion.

"The thinking that got you into the problem", Albert Einstein famously remarked, "is not going to get you out." One cannot expect to change direction if one follows the same driver who is set on moving in one direction – the direction of the path of the *status quo*. The second paper offers a mechanism to change direction by avoiding linear theories. It proposes to replace the mathematical approach, that uses partial differential equations, with the engineering approach. This engineering approach avoids the circular logic that uses material balance in a box to derive partial differential equations that are then discretized into boxes, so the final equations can be linearized! By avoiding this circular approach, the author (the only person to author two books on reservoir simulation) makes the process of reservoir simulation simple and transparent. Now, for once, users of a reservoir simulator will be able to see equations in their raw form and then decide for themselves if and when linearization should be invoked. This paper will launch a new line of reservoir simulation that is bound to revolutionize practices in the petroleum industry. The work finds explanations for every mechanism involved in developing technologies, with particular focus on the oil and gas industries. Considering that this sector is credited as the driver of the modern energy-intensive society, the p;otentiasl impacts and implications are not small.

Even though modeling or visualizing a solution is the most logical start, simulation alone is not enough. One must observe nature and attempt to understand with the hope of emulating the path of nature. This is the only path that is sustainable. Any claim offering sustainability following any other path is patently false. The third paper takes a step forward toward understanding nature according to this framework. It selects fish scale for understanding the ability of nature to absorb toxic chemicals. While the whole world is embroiled in the discussion of how toxic seafood (from fish in the rivers of the North American continent to sharks off the coast of Japan) has become due to contamination from heavy metals, this paper investigates how fish scale (and arguably shark fins, etc.) are made out to be adsorption sites for heavy metals and other toxins. One outcome of this line of work might well be that, one day, we may be able to make use of every bit of fish waste that comes out in order to render the whole process one of zero waste.

No one disputes that we are addicted to our lifestyle, obsessed with tangibles. However, the question arises: when did we start, as a human race, to focus on tangibles so much? The current technology development process, based on the principle, 'the time that matters is now', makes all engineering models forcibly steady state.

This state does not exist, anywhere.

The models based on this steady-state, promoted by Newton and Kelvin, as well as the economist Lord J.M. Keynes, are deeply rooted in Eurocentric culture and promoted as the only natural transition from natural science to engineering. The most important feature of this technology development is the focus on tangibles (time = 'right now'). In medical science, this *modus operandi* amounts to treating the symptom, in economic development, it amounts to increasing wasteful habits to increase GDP, in business, it amounts to maximizing quarterly income even if it means resorting to corruption, in psychology, it means maximizing pleasure and minimizing pain (both in the short-term), in politics, it amounts to obliterating the history of a nation or a society, in mathematics, it means obsessions with numbers and exact (and unique) solutions. In terms of computer, which was initially thought to have modeled human brain, is in fact in stark contradiction with the brain process. With this Science as the only science promoted today, we are often faced with this question: Is this reversible? An analogy of this *modus operandi* lies within medical science, for which it is said: Non-genetic diabetes is reversible by changing the food habit, unless a person is already on insulin. Today's global health is facing the same question: Is global warming reversible? Of course, global warming, much like diabetes, is only a symptom and, like genetic diabetes, there is such a thing as natural warming and cooling that has nothing to do with human activities. But, how do we know if we can do something about this global warming? The fourth paper addresses this question. If we can identify the cause of the global warming and separate the man-made causes from natural ones, we can begin to undo the damage done to lifestyle and reclaim a healthy planet. On this same path of distinguishing man-made from natural causes, near epidemic outbreaks of obesity, diabetes and other conditions improperly ascribed to "living in excessive prosperity" can and should be greatly reduced.

One of the most important features of this journal is that it encourages emulation of nature. It is not a matter of making a claim. It is about actually observing nature and discussing ways to emulate its science. So, what are the most important features of nature? Nature is dynamic (there isn't one phenomenon that is truly steady-state), unique (there isn't one object that is symmetric, homogenous, or uniform), and strictly non-linear (there isn't one entity that is linear, or a process that is linear). That third feature is also the most overwhelming feature of nature. Sure enough, solutions to non-linear problems are also the rarest. In engineering, it is said, engineers love straight lines. This liking may be real, but linearity is not. Just because linear problems are the easiest to solve doesn't mean the solutions are real. There is no justification for forcing linearity as there isn't one straight line in nature. In the Newtonian era, the whole discussion of straight lines boiled down to a dimension that approaches zero. This zero is not the meaningful zero of the oriental culture, it is the zero that means void and aphenomenal. When it comes to the discipline of mathematics, the entire discipline of non-linear mathematics is devoted to simplifying non-linear problems so that the problems are amenable to conventional techniques. Numerical modelers claimed to have solved this problem by introducing matrix solvers after discretizing the governing equations. This indeed added a layer of opacity to the process as the governing partial differential equations are first discretized in their non-linear form only to linearize them prior to solving with the matrix solvers. Of course, matrix solvers can only solve linear algebraic equations. The fifth paper presents the Adomian technique to solve some of the most difficult-to-solve non-linear partial differential equations, as applied in natural waves. The paper proposes no linearization and solves the problem as it appears. This paper will show the way to a new line of research that will avoid linearizing governing equations. Future work should also include determination of all possible solution paths.

All technologies, including those failing us today, carry a "learning-curve" opportunity cost element in their initial development. The argument that simply changing what has been done up to now will incur new, or great, cost does not hold up. The starting point of the technologies discussed in the last paper of this issue is that Nature works best and what is failing us now can be fixed by learning from and following Nature, not by doing more or less with what clearly doesn't work. The solutions – materials, pathways etc. – are all available in existing phenomena of the surrounding natural environment. This is the path of humanizing the environment by putting in command positive intentions to do good, what is right in good conscience, and relying on natural materials and components already at hand as much as possible, rather than industrially-processed substitutes. No, the road to hell is not full of good intentions. Good intentions are instead precondition to perpetually improving our surrounding. Even a bacterium would know this (it certainly operates that way), let alone the best creation of all, namely, humans. The only difference, it seems, is: the others are inherently natural. Humans have to work hard to act on conscience – their natural attribute. Conscience indeed demands that we start any activity with good intentions. This insight is the source of the theme unifying the technologies and science discussed in this article. We have always known that nature operates at zero waste. There isn't one object that is excess in nature, nor there is an object that is not necessary for overall balance. We claim to emulate nature. Yet, when Freon (totally un-natural) was being introduced as an alternative to ammonia (that could be natural), Einstein was granted a patent that would eliminate the need of Freon as well as all moving part of a refrigeration unit. Once more, the corporations and corporatizers looked the other way and ignored this valuable invention of the greatest scientist of modern age and resorted to commercializing Freon-driven refrigeration unit so much that it will take us possibly another century to get rid of the impact. It did take us several decades before we discovered that Freon was behind making the holes in the ozone layer – the layer that protects us. All of a sudden, modifications were proposed to Freon to develop even deadlier and more anti-natural lines of chemicals. Before it takes few more decades to realize we have replaced Freon with something even worse, the authors propose a cooling/heating system that does not need electricity or Freon. In doing so, they improve the measurable global efficiency multifold and brings down the cost to a level that reminds us of the good old days when best things were free. The paper, with its refreshing perspective, takes a significant step forward to develop the link that is needed between tangible or hard science and intangibles.

In: Nature Science and Sustainable Technology
Editor: M. R. Islam, pp. 5-38

ISBN: 978-1-60456-009-1
© 2008 Nova Science Publishers, Inc.

Chapter 2

DEVELOPMENT AND APPLICATION OF CRITERIA FOR TRUE SUSTAINABILITY

M. Ibrahim Khan
Faculty of Engineering, Dalhousie University
Halifax, Canada

ABSTRACT

The concept represented in the terms 'sustainable' or 'sustainability' or 'sustainable development' has been used extensively in the context of environmental, social, economic, and even technological developments. 'Sustainable development' was the key issue discussed in the UN Conference on Environment and Development (UNCED) at Rio de Janeiro in 1992. Sustainable development' has, in terms of the output of that conference, been defined as development that "meets present needs without compromising the ability of future generations to meet their needs". However, 'sustainability' is one of the most abused or misused terms in virtually all contexts. There are numerous definitions of sustainability and not even two definitions have converging meanings. Instead of achieving 'sustainability', some of the sustainable development projects aggravate ecological and other problems, as painfully evidenced in numerous experimental projects.

This paper proposes a new screening criterion for truly sustainable deployment with respect to development projects. The criterion uses time as the only factor to determine sustainability. Because every phenomenon is time dependent, this criterion becomes a powerful screening tool when scientifically calibrated to the demands of specific activities. The development of technologies is evaluated based on this prime criterion, which tells us whether a technology is acceptable or not. Conventional technologies and management tools have been analyzed based on the proposed screening criterion. Different examples are taken from household uses, offshore oil and gas operations, and fish pond and human ecology.

It is suggested that this unique selection criterion is the only means by which to stop present abusive and destructive developments and their future cumulative effects. Moreover, it encourages preserving a healthy natural environment. It can be used to develop inherently sustainable technologies and management tools, and employed by regulatory agencies, process engineers, and resource managers to regulate ecologically sustainable activities and understanding.

Keywords: True sustainability, technology development, sustainability indicators

INTRODUCTION

Few would argue mother's milk is the most natural food for newborn babies. For thousands of years of human civilization, mother's milk had been untainted. Only in the late Modern Age, breast milk is not safer anymore for a baby due to toxic contamination. In many parts of the world, it is reported that mother's milk is contaminated with numerous toxic compounds (Saito et al., 2005). Recent studies reported the presence of high toxic chemicals and carcinogens in breast milk, such as polybrominated diphenylethers (PBDEs), DHBT, hexachlorobenzene (HCB), heptachlor, epoxide and b-hexachlorocyclohexane, polychlorinated dibenzo-p-dioxins (PCDDs), dibenzofurans (PCDFs), biphenyls (PCBs), dichlorodiphenyltrichloroethane and its metabolites (DDTs), hexachlorocyclohexane isomers (HCHs), and chlordane compounds (CHLs) (Lunder. And Sharp, 2003; Cariou et al., 2005; Saito et al., 2005; Kunisue et al., 2006; Sudaryanto et al., 2006).

Thousands years of human history kept the planet a waste-free place without storing any nuclear wastes, non-biodegradable products and toxic compounds. Only recent human activities are causing irreversible damage to the natural world and threaten earth's ability to sustain the future generations. More than sixty percent of world pristine habitats are destroyed or disturbed and species extinction rate is 100-1000 times higher than the normal background rate. In addition, Sea levels will rise, fisheries will collapse, emerging disease epidemics will sweep across the globe and coral reefs will die off and many others to see yet (Millennium Ecosystem Assessment, 2005).

One of the main reasons of these problems is the use of unsustainable technologies. The present trend, especially in technological development, focuses on short-term environmental economic benefits without regard to longer-term consequences. One result is that the technological advances frequently engender severe environmental consequences on an accumulating scale. Some of the consequences are a continuously growing range of pollutants, hazards and types of ecosystem degradation that extend over ever-widening spatial and temporal scales. Recognizing the difficulty, numerous international, national development agencies and NGOs have suggested balancing environmental considerations and economic development. The best-known attempt to draw an ongoing balance between conflicting development imperatives is widely known as 'sustainable development' (WCED, 1987).

As a concept, 'sustainable development' has been used extensively and the practices it requires are considered essential to the future well-being of humanity and the planet. As with other sectors, sustainable development is a hot topic in engineering and technology development, as companies, institutions and authorities compete to develop and implement ever-more sustainable technologies.

New technologies are reported as more (or less) "sustainable" according to an elaborate metrics system (Darton, 2002; Dewulf et al., 2000; Lems *et al.*, 2002; Winterton, 2001). Sustainability metrics for technology assessment are considered as a standard and sought by engineering institutions (e.g. CWRT, 2002; IchemE, 2002), companies (*e.g.*, GlaxoSmithKline (Smith, 2001); Shell (Lange, 2002)) and academics (*e.g.*, Oldenburg University, Delft University of Technology and Ghent University [Dewulf *et al.*, 2000; Eissen and Metzger, 2002; Lems *et al.*, 2002]).

Many different models or frameworks are being used to assess sustainability, but some of the more popular ones are: Global Reporting Initiative (IchemE, 2002), United Nations Commission on Sustainability Development Framework (WCED, 1987), Sustainability Metrics of the Institution of Chemical Engineers (IchemE, 2002) and Wuppertal Sustainability Indicators (Spangenberg and Bonniot, 1998). Generally, these frameworks assess sustainability of development and technology, taking the economic, environmental and social contexts into consideration.

Table 1. Meaning of "sustainable" management or development

Definition of sustainability	Source	Comments
"Development that meets the needs of the present without comprising the ability of future generations to meet their own needs."	WCED (1987)	Most popular definition, but lacks clear direction, i.e., what is the scale of needs?. It can be treated as a 'paper tiger".
"Management practices that will not degrade the exploited system or any adjacent system."	Lubchenco et al. (1991)	Generally system approach. There is no specific direction about time.
"Development without throughput growth beyond environmental carrying capacity and which is socially sustainable."	Daly (1992)	Considers assimilative capacity of nature in a spatial scale. A time direction is completely missing.
"Improvement in the quality of human life within the carrying capacity of supporting ecosystems."	Robinson (1993)	Tries to integrate the social and ecological context in spatial scale, but not in temporal.
"Sustainability is defined as minimizing the consumption of the world's resources by pursuing better environmental performance within product lifecycles."	Donnelly et al., (2006)	Very weak definition; misguided sustainability.

While the approach taken by these conventional models has merit, sight is lost of sustainability of the environment itself, and not just of the specific process or technology: in effect, the main objective of sustainability is not considered. These models suggest achieving sustainability by minimizing risks or damages and remediating problems engendered by the introduction of a given process or technology (Table 1). They do not address the potential problems that could arise from the future introduction of the new technologies, nor do they approach their introduction from the standpoint of not creating new problems in the first place. This latter approach, which aims to establish what might be better described as "truly sustainable" or "inherent sustainability," is the focus of the present paper. The secret of its being achieved lies with time-testing: as $\Delta t \to \infty$, what happens to the process or technology in question?

CONVENTIONAL SUSTAINABILITY

The concept expressed in the terms "sustainable," "sustainability," and "sustainable development" essentially lack clear direction, even though we see them in various government documents, we hear them in mainstream media, and read them in corporate newsletters and international agreements (Wright, 2002). It is also apparent that even those few who may understand the true meaning of sustainability are not able to agree on a criterion (Judes, 2000; Leal Filho, 2000; Wright, 2002). Table 1 shows popular definitions of sustainability.

In the international environmental context, the idea of sustainability is based on the notion that planetary resources are finite, a highly contentious assertion in itself. But essentially, the numerous mutually reinforcing intentional initiatives that have promoted the idea since the Second World War has been united by the need to foster a global understanding of the environment and to address how humanity could ameliorate the delegation of the biosphere.

Founded in the concern for ecological security expressed by the 1972 United Nations Conference on the Human Environment, it was the World Commission on Environment and Development (WCED), that popularized the concept, defining it as the obligation of "meeting the needs of the present without compromising the ability of future generations to meet their own needs" (WCED, 1987). The intervening 15-year period was marked by a titanic battle of ideas in which British Prime Minister Margaret Thatcher, with her declaration that "there is no such thing as 'society'", played a leading role backed by the positions of the Reagan Administration in the United States, which by withholding more than US$1-billion in unpaid dues brought pressure on the FAO and other UN agencies involved in supporting government-sponsored funding of economic development projects among developing countries. Meanwhile, the appeal of the environmental portion of the overall message – the portion that trumpeted "small is beautiful" and similarly idealistic sentiments relating to sustainability – corralled considerable support among younger academics and other young people, and thus it came about that sentiments similar to those of the Brundtland Report were echoed in the UNCED, the United Nations Conference on Environment and Development, held in Rio de Janeiro, Brazil in June 1992.

Criticisms of the Brundtland definition of sustainable development exist (Table 1), including those centered on the anthropocentric connotations of the wording; that it generates strategies that ignore the carrying capacity of the planet; that it contains vague wording which allows individuals to manipulate the definition; and that it apparently supports sustaining growth (Brown *et al.* 1991; Leal Filho, 2000; Gibson, 1991; Hawken, 1992; Miller 1994; Nikiforuk 1990; Rees 1989; Wackenagel and Rees, 1996; Welford 1995). Recently, Appleton (2006) criticized that Brundtland definition of 'satisfying human need', but who much is a limit for human. He argued that what level human needs should be satisfied. Is it the American per capita income level, the Chinese per capita income level, the Millennium Development Goals or some similar bundle of minimum services (i.e. clean water, healthy sanitary conditions, an x calorie and nutrient a day diet, and heat levels in the winter)? He also questions, such as is air conditioning in the summer a human need; is the measurement of need ownership of 0.6 cars per capita; as in the United States, or one car per capita, the current worldwide ratio, or the current 0.08 cars per capita in China? However, this definition is very popular to date and has been used in many policy and government documents worldwide. Someone also argued that this definition is weak and based on perfect direction.

To assess the sustainability of projects, technologies as well as the overall implications of different types of 'sustainability frameworks' has been used (Labuschagne et al., 2005). Some of them are: Global Reporting Initiative (GRI, 2002), United Nations Commission on Sustainability Development Framework (UNCSD, 2001), Sustainability Metrics of the Institution of Chemical Engineers (IChemE, 2002) and Wuppertal Sustainability Indicators (Spangenberg and Bonniot, 1998).

The Global Reporting Initiative (GRI) was launched in 1997. The United Nations Environment Program together with the United States NGO, Coalition for Environmentally Responsible Economics (CERES), developed GRI. The main goal of GRI is enhancing the quality, rigor and utility of sustainability reporting (GRI, 2002). Reporting is, therefore, the focal point of these guidelines. The GRI uses a hierarchical framework in the three focus areas, namely, social, economic, and environmental.

The United Nations Commission on Sustainable Development (UNCSD) constructed a sustainability indicator framework for the evaluation of governmental progress in achieving sustainable development goals. A hierarchical framework groups indicators into 38 subthemes and 15 main themes that are divided between the four aspects of sustainable development (UNCSD, 2001). The main deference between this framework and the GRI, for example, is the fact that it addresses institutional aspects of sustainability.

The Wuppertal Institute proposed indicators for the four dimensions of sustainable development, as defined by the United Nations CSD, together with interlinkage indicators between these dimensions (Spangenberg and Bonniot, 1998). These indicators are applicable at both macro and micro levels. The approach used to address business social sustainability is worth noting: the UNDP Human Development Index has been adapted to form a Corporate Human Development Index.

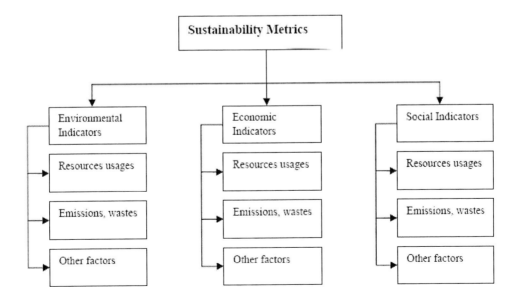

Figure 1. Basic components of sustainability metrics.

The Institute of Chemical Engineers (IChemE) published a set of sustainability indicators in 2002 to measure the sustainability of operations within the process industry (Figure 1). The IChemE provides standard reporting forms and conversion tables. This framework is less complex and impact oriented. However, the framework strongly favors environmental aspects, as well as quantifiable indicators that may not be practical in all operational practices, *e.g.*, in the early phases of a project's life cycle.

A review of the literature that has sprung up around the concept of sustainability indicates, however, a lack of consistency in its interpretation. More importantly, while the all-encompassing nature of the concept gives it political strength, its current formulation by the mainstream of sustainable development thinking contains significant weaknesses (Wright, 2002). Most of the sustainability assessment models, which are mentioned above, are based on broader socio-economic and environmental considerations. Overall, even though it is obvious that different matrix systems and indexes have been used to measure sustainability, there are no straightforward guidelines to achieve true/inherent sustainability.

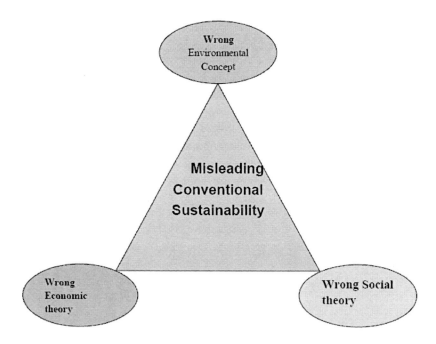

Figure 2. Conventional Sustainability based on wrong economic, environmental and social theory

The problem with currently developed technologies appears almost daily in the form of 'overlooked' features of a new product, yet none of the criteria proposed to-date could not have screened out these technologies (Table 2). In fact, all of them received clearing from regulatory agencies and none seems to be concerned that the 'recall' rate is high.

TRUE SUSTAINABILITY

The term 'sustainability' and 'sustainable development' have become the main focus of debate after the UN convention on 'Environment and Development' in 1979 (Wright, 2002;

Khan and Islam, 2005a; Khan et al., 2005 and 2006a; Appleton, 2006). From then on, this term 'sustainability' has been referred to all sectors, such as technology development, business operation, natural resource management and others. However, it is proven that the sustainable development became a 'paper tiger' with respect to achieving desired progress or stopping acute toxic impacts of technology development. The main reason behind this failure is that the present sustainability concept is formulated based on "wrong theory". Figure 2 pictorially shows the wrong concepts of conventional sustainability. It is obvious that if a theory is based on wrong principle then it cannot achieve correct result. For example, if the economic theory is wrong then how could it be possible to achieve economic progress?

Wrong Economic Theory

Brian Diliman (Pharmacist), said, "It seems economics is always tied to change". He was referring to why health products are constantly changing focus, touting one brand of product today and discarding them all the next day for yet another brand of products. Here again, the key word is 'change' – there is only one kind and it can only make things worse for the general population while amassing benefit for a select few who are better off invoking the change at a non-stop pace. What is entropy in Physics, 'change' is in economics. Just like entropy is thought to be increasing disorder, 'changes' are thought to create confusion, hence, more money in the pocket of those who benefit from the 'disorder'. For this, the more changes one can invoke, with more obsessions for the short-term, better off one's corporation would be. This focus on short-term is extended to 'Modern Economic" theory by Lord J.M. Keynes, who said, "In the long run, we are all dead" (Zatzman and Islam, 2006). This is indeed a natural extension of Freud's Psychoanalysis to Economics. This would be merely the beginning of the corporate culture that keeps on producing one scandal after another.

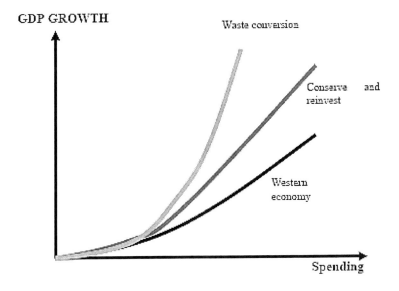

Figure 3. The GDP growth analysis gives distorted view of the economic outlook and limits options for improvement (redrawn from Islam, 2005a)

Because of the top-down model, many baseless schemes were introduced. The economic sustainability or success of a country is widely measured by the gross domestic products (GDP). According to the present trend, the GDP is positively correlated to economic progress. If GDP index increase in any country, then the economic growth is assured. However, in-depth analysis shows that the higher index of GDP does not represent the real progress of a country. This is shown in Figure 3. The current economic theory does not differentiate between spending money on education or wasteful, self-destructive activities. Consequently, the current model has no room for promoting sustainable investments (Zatzman and Islam, 2006). Figure 3 demonstrates that by simply differentiating among random investment that promotes wasteful habits, investment in waste minimizing, and investment in waste conversion (into value added products), one would be able to have a wide range of GDP growth for the same amount of investment. According to the aggregate expenditure model in macroeconomics the GDP can be defined as Equation 1.

$$GDP = C + I + X_n + G \qquad 1$$

Where,
  C = consumption expenditure
  I = Investment
  X_n = Net export, X_n = X-M
  X = export
  M = Import
  G = Government expenditure
  GDP = Gross Domestic Production

According to this model, investment expenditure and GDP are positively related. That means an increase in Investment leads to an increase in GDP. And how much GDP will increase depends on the investment MULTIPLIER. Investment Multiplier is the ratio of a change in equilibrium GDP to the change in investment.

$$\text{Investment Multiplier} = \frac{\text{Change in GDP}}{\text{Change in Investment}} \qquad 2$$

Usually, this index is greater than unity. It means that the change in GDP is greater than the change in investment. Therefore, according to the economist, a small increase in investment will lead to increasing greater increase in GDP.

$$GDP = C\uparrow + I\uparrow + X_n\uparrow + G \qquad 3$$

Our observation is that the increase in investment does not increase the GDP or the overall economic status of a country. In Figure 3, the represents the increase of components. More importantly, unjustified investment is likely to worse the economic condition. There a number of examples discussed in the case of natural resource management.

Fisheries Sector

Above mentioned traditional economic formula does not function in the fisheries sector. For example, we can imagine that a maritime nation suddenly increases the investment in the fisheries sector. They bought 100 fishing vessels at the cost of a billion dollars. Using these modern gears, all commercial fish are harvested. According to this economic formula, the GDP of this small nation increased significantly. However, due to over fishing, there was no brood fish left for breeding to increase the fish population. Then more fishing vessels would not lead to higher yield if the resources are depleted. The higher investment in fishing industry will be further depleted the fisheries stock and destroy its ecological balance. Finally, fisheries are collapsed to ruins. Considering this scenario, one can say that more investment might not necessarily lead to improving the economy rather it may destroy the economy. This analysis only included tangible or short-term costs. If intangibles, such as long-term cost of damaging the ecosystem, are included, the 'bigger the better' principle implodes even more spectacularly.

Tobacco Sector

More investment in the tobacco sector means higher production, marketing, and consumption of cigarettes. According to conventional approach it will increase the GDP and it might increase short-term government revenue. However, higher cigarette consumptions will lead to more respiratory and other tobacco relegated diseases, because, cigarette smoke produces many know and unknown toxic compounds (Islam, 2003). A typical sign in a Halifax grocery store says: "tobacco kills – Tobacco contains an addictive drug and kills 1400 in Nova Scotia province of Canada, each year - more than accidents, alcohol, AIDS, homicides, and suicides combined" (Islam, 2003). This killing is just intended of making money and cigarette companies never get punished. It is estimated each human possesses $45 million-worth of usable body parts. It is as a sum of $63 billion. This amount is never considered in GDP estimations (Islam, 2003).

According to present economic model, if spending is increased, GDP will increase, irrespective of nature of spending. This is such that GDP growth is deemed ipso facto "good" regardless of whether the increased amount of exchangeable value was due to expanded weapons production and exports, or to increased consumption of life-shortening quantities of alcohol, tobacco, or anything else refined to truly toxic levels (Islam, 2005b). A new approach of GDP is shown in Figure 4. The good (by term) investment is shown as positive value and bad (short-term) investment will be considered as negative value to estimate GDP. As a result, the GDP might be positive (+) or negative (-).

The metric of money gained or lost is not only grossly linearized, such a metric is also utterly external to and indifferent about whether a given economic activity served any real need of Humanity in the long term or the short term. In this respect, such a metric is indeed aphenomenal beyond $t =$ 'right now.' It is very easily conceived that spending on long-term projects should fetch much greater benefit than short-term projects. However, modern economic theory made no provision for that analysis (Zatzman and Islam, 2006).

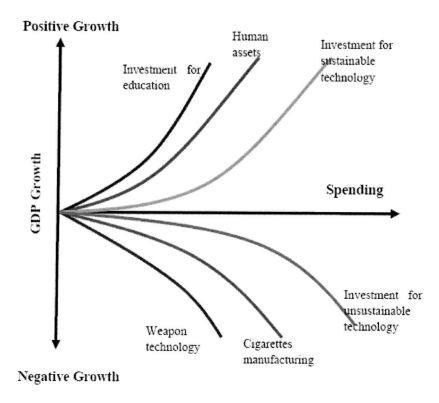

Figure 4. Real representation of GDP

These examples show that present economic theories are misleading and do not address sustainability. Therefore, to achieve true economic progress a new economic theory is needed. There is also big gap between ecology and economics. For the real progress the ecology and economics should work together. Similarly, if the environmental concept is wrong, then how would it be possible to achieve sustainable environmental development. How could we develop a sustainable technology, if the technological development guidelines are wrong? Mittelstaedt (2006) there are over 40000 toxic chemicals in daily use. Only recently, United Arab Emirates banned the use of plastic containers for serving hot beverages (Gulf News, June 20, 2006). It is only matter of time that plastic will in eve to be banned even for cold food. However, while technological development foresees this grim picture? None. In the present development regime, there are misleading concepts in economic, environmental and social sectors. This grim scenario, however, can be avoided through introducing true sustainability which will be based on true principles/concepts of sustainability.

Problem with Social Theory

For the entire history of the Industrial Revolution – beginning in Britain in the middle of the 18th century – one constant theme has been the ever-increasing scale on which production carries on. The drive towards more cost-effective energy systems powering industry – from coal to petroleum to electricity to natural gas – has encouraged ever greater levels of concentration of production, while the ruinous consequences of periodic crises of overproduction led stronger capital formations to absorb weaker formations, thereby – over time – increasingly concentrating ownership of the productive forces as well. The combined concentration of production itself with concentration of its ownership has accelerated tendencies towards integration of primary, secondary and tertiary phases of production and distribution under a single corporate umbrella, in a process generally labeled "vertical integration." The notion of "transmission belt" provides a unifying common metaphor defining the interrelationships of the various structures and functions under such an umbrella. According to this metaphor, demand for products on the one hand and their supply on the other hand can be balanced overall, even though certain markets and portions of the corporate space may experience temporary surpluses or shortages of supply or demand. This apparently elegant unity of structure, function and purpose at the level of very large-scale monopolies and cartels is one side of the coin. The other side is seen in the actual fate of the millions of much smaller enterprises. For these entities, partial integration down to outright hard-scrabble competitive scramble-for-position is the rule, and they remain always much more vulnerable to mismatching of and swings in demand and supply. The conventional accounting for these phenomena assumes that the vertically-integrated "upstream-downstream" model is the most desirable and the fate of the small enterprise is some throwback that "economic evolution" will dispose of in the same way that entire species disappear from the natural world of the plant and animal kingdom. What the conventional account fails to explain, however, is the utterly unsustainable scale of resource rape and plunder that the assertedly superior vertically-integrated "upstream-downstream" enterprise seems to require in order to maintain its superiority.

From de Quesnay's *Tableau économique* in the 1760s to the work of 1973 Nobel Economics laureate and Harvard University professor Wassily Leontieff *et al.* in the 1930s, 1940s and 1950s (Leontieff 1973), transmission-belt theories of productive organization, along with all the "input-output" models of economic processes spawned from its premises, have focused entirely, and actually very narrowly, on accounting for the circulation of capital – and nothing else. During the 19th century Karl Marx's work on the circulation of capital (Marx 1883), and in the 20th century the works produced by the neo-Ricardian school of classical economics developed by Piero Sraffa *et al.*, starting in the 1920s and culminating in his *Production of Commodities By Means of Commodities* (Sraffa 1960) demonstrated that the rich getting richer and the poor getting poorer was inherent and inevitable in any system that buys and expends the special commodity known as labor-time in order to produce whatever society needs in the form of commodities. Their research established that the circulation of variable capital (wages spent to buy consumer goods) runs counter / opposite to the circulation of constant capital (money spent on raw materials used in production and for replacing, repairing or maintaining machine and equipment used up or partially "consumed" in the processes of production). In effect: merely redistributing the output of such a system more equitably cannot overcome inherent tendencies of such systems towards either crises of overproduction or a cumulative unevenness of development with wealth accumulating rapidly at one pole and poverty at the other. These tendencies can only be overcome in conditions where people's work is no longer treated as a commodity.

There are various pathways to decommodifying labor-time. For an economics of intangibles, time is considered as the quantity of labor-time needed to produce / reproduce society's needs on a self-sustaining basis. Supply is planned on the basis of researched knowledge. Its provision is planned according to the expressed needs and decision of collectives. Meanwhile, production is organized to maximize inputs from natural sources on a sustainable basis. Primary producers organize through their collectives. Industrial workers and managers organize their participation in production through their collectives. Those involved in wholesale or retail distribution organize their participation through their collectives. The common intention is not the maximum capture of surplus by this or that part of the production-distribution transmission-belt, but rather to meet the overall social need.

One of the greatest obstacles to achieving sustainable development on these lines is that the pressure to produce on the largest possible scale. Although this is intended to realize the greatest so-called "economies" and "savings of labor", it also comes into sharp contradiction with the competitive drive by individual enterprises to maximize returns on investment in the short term. Modern society grants all intentions full freedom to diverge, without prioritizing the provisioning of social needs. This means, however, that every individual, group or enterprise considers its cycle and social responsibility completed and discharged when the short-term objective, whatever it may be, is achieved. The only way to engender and spread truly sustainable development, on the other hand, is to engage all individuals and their collectives for, and according to, common long-term intentions and, on that basis, harmonize the differing interests in the short-term of the individual and the collective.

Although the execution of any plan, however well-designed, requires hierarchy, order and sequence, it does not follow at all that plans in themselves must also be produced in a top-down manner. This top-down approach to drawing up economic plans was one of the gravest weaknesses of the eastern-bloc economies that disappeared with the Soviet Union. Starved of the flow of externally-supplied credit, the bureaucratized centre on which everything depended became brain-dead. Nowhere else in their systems could economic initiative be restored "from below." The Republic of Cuba learned this lesson the hard way when Soviet support disappeared and their economic output shrank by more than one-third. It was not just the extreme and government-enforced rationing measures of the subsequent Special Period, however, that saved the day. It was that these measures were drawn up from and at the base, and they included the extremely sage decision not to close a single clinic or medical facility or school or educational facility, whatever else might be sacrificed in the short term. Cuba today is the world's most advanced exporter of health and education services – including a wide range of specialized emergency hospital and clinic setups, effective but non-commercial drugs and medicines, and even 90-day crash literacy courses in Spanish and some other languages.

The destructiveness of the U.S. prejudices about Cuba, meanwhile, has started taking its toll of American lives. In the summer of 2005, the world watched with astonishment as a season of severe hurricanes lashed the entire Gulf of Mexico, including the Gulf Coast of the United States, the island of Cuba, the Yucatan Peninsula in Mexico, etc. Two of the most destructive hurricanes to hit Cuba directly (one of them, Rita, went on to lash the US Gulf Coast as well) forced the temporary evacuation of more than 2.5 million residents, but not a single person lost their life nor was there any report of looting of personal property. This was not surprising news around the Caribbean, where Cuba has acquired an international reputation in the field of hurricane emergency response. Meanwhile, however, the situation unfolding on the U.S. Gulf Coast must rank as one of the great tragedies of an already severely ravaged 21st century. Hurricanes Rita and, earlier, Katrina, created a huge disaster in the city of New Orleans, rendering much of it uninhabitable after the long-neglected levee system gave way. With the publication (on 1 March 2006) of the evidence of a video conference call involving President George W Bush and senior federal and state emergency management officials the day before Katrina struck, it has become established as fact that the highest levels of the U.S. government took no serious preventive action in advance of the expected levee breach. At the same time, only days after Katrina, the U.S. government rejected a sincere offer by Cuba of more than 1,500 emergency doctors and other trained medical personnel. Meanwhile (as of March 2006), there were still no reliable estimate of the final death toll or total of the numbers of New Orleans residents displaced by and since the hurricane. It is becoming complicated to sort out which is the greater crime against humanity: the U.S. government leaving one of its cities to drown in its own sewage, or a prejudice so reactionary that tens of thousands of Gulf Coast residents were left without any medical assistance at a most critical moment?

Could the Cubans' notion of saving, preserving and enhancing human capital as a fundamental requirement for truly sustainable development be implemented in such directions if they had relied exclusively on top-down methods? The answer must be: No. Having decided to facilitate the fullest possible expression of a humanitarian intention, they proceeded to accomplish the aim as expeditiously as possible. This lesson is being applied repeatedly in all other areas of Cuban life. For example, as a result of the continuing U.S. trade embargo combined with the loss of food imports from the former Soviet bloc, Havana today is the only city on the planet that is approaching self-sufficiency in the provision, from rooftop gardens and plots on vacant land around that city, of fresh and entirely organic fruits and vegetables for its nearly two million residents – something unthinkable without extensive planning developed from the base.

Truly sustainable development is probably not possible under the current legal systems of North America. In this arena, the greatest obstacle arises from the law's prioritizing of protection for property that can be used to make money or exploit others over protection of people's individual and collective rights as individuals born to society. One of the forms in which this finds expression is the definition and treatment of corporate entities as legal persons, who unlike physical persons are exempt from imprisonment upon prosecution and conviction for the commission of criminal acts. The prioritizing bias of the existing legal system, expressed most rigidly in its unconditional defense of the sanctity of contract, has many implications as well for the manner in which money and credit may be offered or withdrawn in conventional business dealings among business entities, corporate or individual: the shifting needs of real human collectives affected by such changes take second place to the actual wording of a contract, regardless of the conditions in which it was drawn up. This gives wide latitude for any individual or group that would seek to wreck the interests of any well-intended collective, while affording the victims of sharp practice almost no remedy.

Problem with Technology Development

The technologies promoted in the post-industrial revolution are based on the aphenomenal model. This model is a gross linearization of nature ("nature" in this context includes Humanity in its social nature). This model assumes that whatever appears at $\Delta t = 0$ (or time = 'right now') represents the actual phenomenon. This is clearly an absurdity. How can there be such a thing as a natural phenomenon without a characteristic duration and-or frequency? When it comes to what defines a phenomenon as truly natural, time in one form or another is of the essence.

The essence of the modern technology development scheme is the use of linearization or reduction of dimensions in all applications. Linearization has provided a powerful set of techniques for solving equations generated from mathematical representations of observed physical laws – physical laws that were adduced correctly, and whose mathematical representations, as a symbolic algebra, have proven frequently illustrative, meaningful and often highly suggestive. However, linearization has made the solutions inherently incorrect. This is because, any solution for t = 'right now' represents the image of the real solution, which is inherently opposite to the original solution. Because this model does not have a real basis, any approach that focuses on short-term makes on travel on the wrong path. Unlike common perception, this path does not intersect the true path at any point in time other than t ='right now'. The divergence begins right from the beginning. Any natural phenomenon or product always travels an irreversible pathway that is never emulated by the currently used aphenomenal model of technology development. Because, by definition nature is non-linear and 'chaotic' (Gelieck, 1987), any linearized model merely represents the image of nature at a time, t = 'right now', from which instant their pathways diverge. It is safe to state all modern engineering solutions (all are linearized) are anti-nature. Accordingly, the black box was created for every technology promoted (Figure 5).

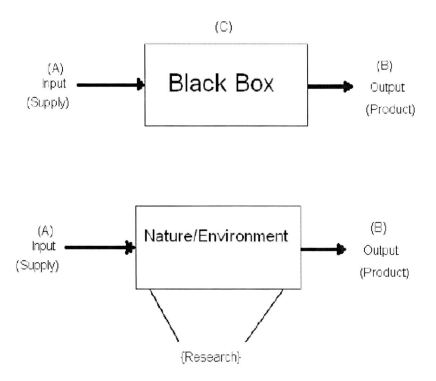

Figure 5. Classical "engineering" (top) and 'Nature-based' notion (bottom) (redrawn from Islam, 2005a)

This formulation of a black box helped keep "outsiders" ignorant of the linearization process that produced spurious solutions for every problem solved. Its crucial and least remarked feature? The model itself has nothing to do with knowledge. In a typical repetitive mode, the output (B) is modified by adjusting input (A). The input itself is modified by redirecting (B). This is the essence of the so-called feedback mode that has become very popular in our day. Even in this mode, nonlinearity may arise as recently efforts have been made to include a real object in the black box. This nonlinearity is expected as even a man-made machine would generate chaotic behavior, which becomes evident only if we have the means of detecting changes over the dominant frequency range of the operation.

What needs to be done is to improve knowledge of the process. Before claiming to emulate nature, we must implement a process that allows us to observe nature (Figure 5). Research based on observing nature is the only way to avoid spurious solutions due to linearization or elimination of a dimension.

Sustainable development is characterized by certain criteria. The 'time' criterion is the main factor in achieving sustainability in technological development. However, in the present definition of sustainability a clear time direction is missing (Table 1). To better understand sustainability, we can say that there is only one alternative to sustainability, *viz.*, unsustainability. Unsustainability involves a time dimension: it rarely implies an immediate existential threat. Existence is threatened only in the distant future, perhaps too far away to be properly recognized. Even if a threat is understood, it may not cause much concern now, but will cumulatively work in its effect in the wider time scale. This problem is depicted in Figure 6.

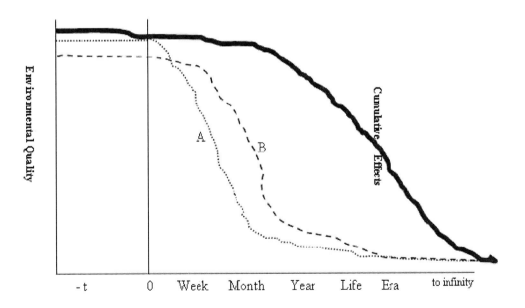

Figure 6. Cumulative Effects of Activities A and B within different Temporal Periods.

In Figure 6, the impact of the wider time scale is shown where 'A' and 'B' are two different development activities that are undertaken in a certain time period. According to the conventional environmental impact assessment (EIA) or sustainability assessment process, each project has insignificant impacts in the environment in the short time scale. However, their cumulative impacts will be much higher and will continue under a longer time scale. The cumulative impacts of these two activities (A, B) are shown as a dark line.

Technology plays a vital role in the modern society. The causes of present-day environmental and social problems are related to the use of unsustainable technology. The problem associated with the current technology development mode is easily identified if one attempts to correlate between the increase of cancer, non-genetic diabetes, Alzheimer's disease, and others, with the increase in the usage of sugar ('refined' biofuel), gasoline ("refined" fossil fuel), and petrochemicals (ranging from fertilizers to ubiquitous plastics materials). This has recently been highlighted by Islam (2005c). Most modern technologies are developed on principles that focus on short-term economic benefits.

The most important feature of this technology development is the focus on tangibles. The overwhelming assumption behind this kind of 'technology development' is that Nature is chaotic and needs fixing. The most important manifestation is the fascination with imitations, homogenization, and plastics. Today, we use 90 million barrels of crude oil and produce 4 million tonnes of plastic everyday. Yet, reduce, reuse, and recycle remains the only slogan available today. Disinformation is the essence of the current technology development scheme (Table 2).

Table 2. Analysis of "breakthrough" technologies

Product	Promise (knowledge at t= 'right now'	Current knowledge (closer to reality)
Microwave oven	Instant cooking (bursting with nutrition)	97% of the nutrients destroyed; produces dioxin from baby bottles
Fluorescent light (white light)	Simulates the sunlight and can eliminate 'cabin fever'	Used for torturing people, causes severe depression
Prozac (the wonder drug)	80% effective in reducing depression	Increases suicidal behavior
Anti-oxidants	Reduces aging symptoms	Gives lung cancer
Vioxx	Best drug for arthritis pain, no side effect	Increases the chance of cancer
Coke	Refreshing, revitalizing	Dehydrates; used as a pesticide in India
Transfat	Should replace saturated fats, incl. high-fiber diets	Primary source of obesity and asthma
Simulated wood, plastic gloss	Improve the appearance of wood	Contains formaldehyde that causes Alzheimer
Cell phone	Empowers, keep connected	Gives brain cancer, decreases sperm count among men.
Chemical hair colors	Keeps young, gives appeal	Gives skin cancer
Chemical fertilizer	Increases crop yield, makes soil fertile	Harmful crop; soil damaged
Chocolate and 'refined' sweets	Increases human body volume, increasing appeal	Increases obesity epidemic and related diseases
Pesticides, MTBE	Improves performance	Damages the ecosystem
Desalination	Purifies water	Necessary minerals removed
Wood paint/varnish	Improves durability	Numerous toxic chemicals released
Leather technology	Won't wrinkle, more durable	Toxic chemicals
Freon, aerosol, etc.	Replaced ammonia that was 'corrosive'	Global harms immeasurable and should be discarded

Source: Islam, 2005a and 2005b

The present-day technological development process focuses on turning maximum profit in minimum time with expenses of many environmental consequences. As human beings in a fast-moving modern society, our vision of time is extremely short-term. For example, we commonly think in terms of a bi-weekly pay cheque. This two-week period is our standard in the civilized world and it has its reflection in development activities. Long-term planning may be considered within time frames ranging from hours to weeks, months or perhaps even years – but never in terms of, say, generations (25-year periods). This short-term focus is exactly opposite to what is needed to ensure sustainability.

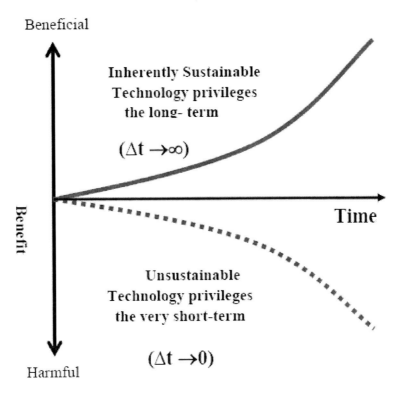

Figure 7. Direction of sustainable and unsustainable technology

SUSTAINABILITY IN TECHNOLOGY DEVELOPMENT

Sustainability can be assessed only if technology emulates nature. In nature, all functions or techniques are inherently sustainable, efficient and functional for an unlimited time period. In other words, as far as natural processes are concerned, *'time tends to Infinity'*. This can be expressed as t or, for that matter, $\Delta t \to \infty$.

By following the same path as the functions inherent in nature, an inherently sustainable technology can be developed (Khan and Islam, 2005b). The 'time criterion' is a defining factor in the sustainability and virtually infinite durability of natural functions. Figure 7 shows the direction of nature-based, inherently sustainable technology, as contrasted with an unsustainable technology. The path of sustainable technology is its long-term durability and environmentally wholesome impact, while unsustainable technology is marked by Δt

approaching 0. Presently, the most commonly used theme in technology development is to select technologies that are good for t='right now', or $\Delta t = 0$. In reality, such models are devoid of any real basis (termed "aphenomenal" by Khan et al., 2005) and should not be applied in technology development if we seek sustainability for economic, social and environmental purposes.

Considering pure time, so to speak (or *time tending to Infinity*) in terms of sustainable technology development raises thorny ethical questions. This 'time tested' technology will be good for Nature and good for human beings. The main principle of this technology will be to work towards, rather than against, natural processes. It would not work against nature or ecological functions. All natural ecological functions are truly sustainable in this long-term sense. We can take a simple example of an ecosystem technology (natural ecological function) to understand how it is time-tested (Figure 8).

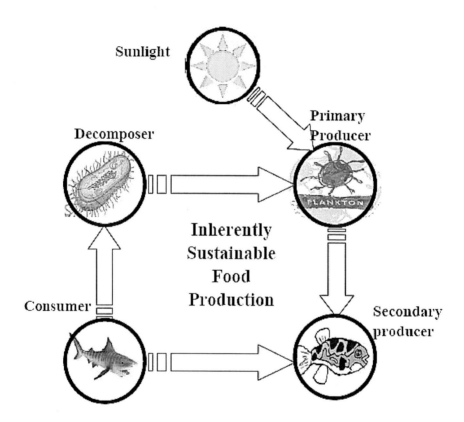

Figure 8. Inherently sustainable natural food production cycle

In nature, all plants produce glucose (organic energy) through utilizing sunlight, CO_2 and soil nutrients. This organic energy is then transferred to the next higher level of organisms which are small animals (zooplankton). The next higher (tropical) level organism (high predators) utilize that energy. After the death of all organisms, their body mass decomposes into soil nutrients which again take plants to keep the organic energy looping alive (Figure 8). This natural production process never dysfunctions and remains for an infinite time. It can be defined as a time tested technique.

This time-tested concept can equally apply in technology development. The new technology should be functional for an infinite time. This is the only way it can achieve true sustainability (Figure 9). This is the idea that informs the new assessment framework that is developed and shown in Figure 9 and 2.10. The triangular sign of sustainability in Figure 9 is considered as the most stable sign. In this a triangle is formed by different criteria which represent a stable sustainability in technology development. This is the idea that informs the new assessment framework that is developed and shown in Figure 10. Any new technology could be evaluated and assessed by using this model. There are two selection levels. One is the primary level and other a secondary level. A technology must fulfill primary selection criterion before being taken to the secondary level of selection. The primary selection criterion is "time".

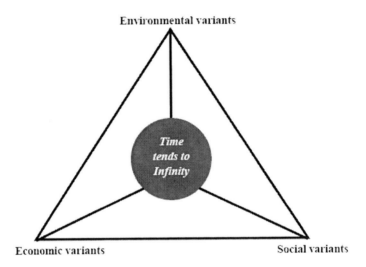

Figure 9. Pictorial view of the major elements of sustainability in technology development.

For the sake of simulation, we imagine that a new technology is developed to produce a product name "ever-rigid". This product is non-corrosive, non-destructive and highly durable. The 'ever-rigid' technology can be tested using the proposed model to determine whether it is truly sustainable or not. The prime first step of the model is to find out if "ever-rigid" technology is "time-tested." If the technology is not durable over infinite time, it is rejected as an unsustainable technology and would not be considered for follow-up testing. For, according to the model, time is the prime criterion for the selection of any technology.

If the "ever-rigid" technology is acceptable with respect to this time criterion, then it may be taken through the next sorting process to be assessed according to a set of secondary criteria. The initial set of secondary criteria analyzes environmental variants. If it passes this stage, it goes to the next step. If this technology is not acceptable with respect to environmental factors, then it might be rejected, or further improvements might be suggested to its design. After environmental evaluation, the next two steps involve technological, economic and societal variants analyses, each of which follows a pathway similar to that used to assess environmental suitability. At these stages also, either it will ask for improvements required to be made to the technology, or it might be rejected as unsustainable.

24   M. I. Khan

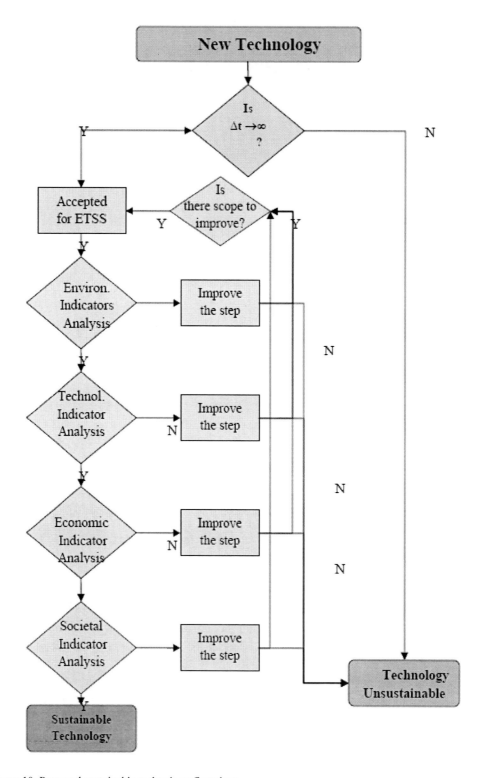

Figure 10. Proposed sustainable technology flowchart

TOOLS TO EVALUATE SUSTAINABILITY

Sustainability System Approach

In order to consider a technology that is inherently sustainable a method is needed for evaluation. This evaluation method should be based on principle of true sustainability which is defined and shown in the form of flowchart in Figure 10.

Based on this newly developed method a practical tool is proposed and presented shown in Figure 11. In this evaluation method, for the shake of sustainability the total critical natural resources should be conserved in a whole technological process. As well, waste produced in the process of using the technology should be within the assimilative capacity of a likely to be affected ecosystem. This means that an intra and intergeneration ownership equity of natural resources on which the technology depends must be ascertained (Daly, 1999, Costonza et al., 1997). Daly (1999) points out that all inputs to an economic process, such as the use of energy, water, air, etc., are from natural ecology and all the wastes of the economic process are sunk back into it. In other words, energy from the ecological system is used as throughput in the economic process, and emissions from the process are given back to ecology. In this sense, an economic system is a subsystem of ecology and, therefore, the total natural capital should be constant or increasing. Man-made capital and environmental capitals are complementary but are not substitutable. As such, any energy system should be considered sustainable only if it is socially responsible, economically attractive and environmentally healthy (Islam, 2005c).

To consider any energy system sustainable, it should fulfill basic criteria with respect to environment, social, economic and technology (Pokharel et al., 2003 and 2006; Khan and Islam, 2005b and 2005c; Khan et al., 2005 and 2007). In this study following criteria are taken into consideration:

Natural (environment) Capital (Cn) + Economic Capital (Ce) + Social Capital (Cs) ≥ Constant for All Time Horizons

$(C_n + C_e + C_c)_t \geq$ constant for any time 't' provided that $\dfrac{dCn_t}{dt} \geq 0$, $\dfrac{dCe_t}{dt} \geq 0$, $\dfrac{dCs_t}{dt} \geq 0$. These conditions are shown in a flow chart format in Figure 11. In the proposed model, a technology is only 'truly sustainable' if it fulfills the time criterion. Other important criteria that it must also fulfill are related to environmental, social and economic factors as shown in 2.12. In the broader view, therefore, sustainable development in terms of technological development, is seen as having four elements namely, economic, social, environmental, technological, cost effectiveness and even regulatory requirements (Figure 13, 14, 15 and 16).

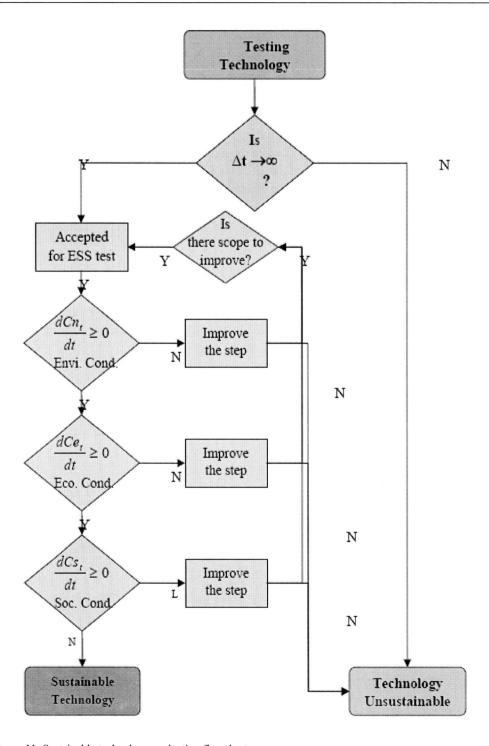

Figure 11. Sustainable technology evaluation flowchart

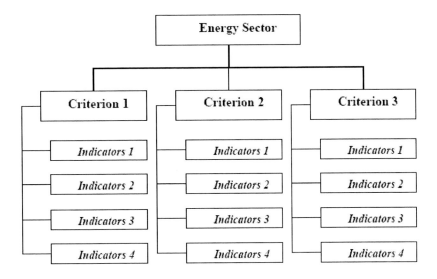

Figure 12. Hierarchical position of criteria and indicators of sustainability

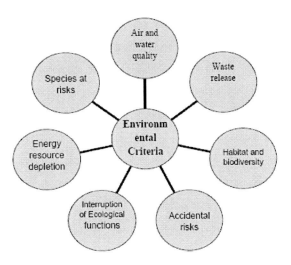

Figure 13. Environmental criteria

Indicators for Sustainable Development

Sustainable development is accepted as a vision for managing the interaction between the natural environment and social and economic progress with respect to time. However, there is no suitable method to measure it and experts are still struggling with the practical problem of how to measure it. The Centre d'Estudis D'Informaci Ambiental (CEIA, 2001) stated that "the move towards sustainability would entail minimizing the use of energy and resources by maximizing the use of information and knowledge". In effect, in order to develop sustainable technology and manage natural resources in a sustainable manner, decision-and policy-makers need to improve the application of knowledge gained form information. However, there is generally a large communication gap between the provision of data and the application of knowledge.

One method of providing information in a format that is usable by policy- and decision-makers is through the use of sustainability indicators. An indicator is a parameter that provides information about and environmental issue with a significance that extends beyond the parameter itself (OECD 1993 and 1998). Indicators have been used for many years by social scientist and economists to explain economic trends, a typical example being Gross National Product (GNP). Different NGOs, government agencies and other organizations are using indicators for addressing sustainable development. Some of them are: the World Resources Institute; the world conservation Union-IUCN; United Nations Environmental Program; the UN Commission on Sustainable Development; European Environmental Agency; the International Institute of Sustainable Development (IISD); the World Bank (IChemE, 2002).

Figure 14. Community criteria to consider for sustainability study of offshore oil and gas

Indicators for addressing sustainable are widely accepted by development agencies in national and international level. For example, *Agenda 21* (Paper 40) states that "indicators of sustainable development need to be developed to provide solid bases for decision-making at all levels and to contribute to the self-regulating sustainability of integrating environmental and development systems" (WCED, 1987). This has led to the acceptance of sustainability indicators as basic tools for facilitation public choices and supporting policy implementation (Dewulf and Langenhove, 2004; Adrianto et al., 2005). It is important to select suitable indicators, because they need to provide information on relevant issues; identify development-potential problems and perspective; analyses and interpret potential conflicts and synergies, and assist in assessing policy implementations and impacts.

Figure 15. Policy criteria to consider for the sustainability of offshore oil and gas

Khan and Islam (2005a and 2006a) developed sets of indicators for technology development and offshore oil and gas operations. The hierarchical positions of criteria and indicators are presented in Figure 12. They developed indicators for environmental, societal, policy, community and technological variant which are shown in Figure 13, 14, 15 and 16. Analyzing these sets of indicators they also evaluated the sustainable state of offshore operation and its technology. Figure 17 shows the sustainability state of different offshore seismic technology. In the figure 4D, 3D and nature-based technology got the higher score and explosive got the lowest score. Outer boundary of the web represents the targeted sustainability, but none of the existing technologies got that level of score.

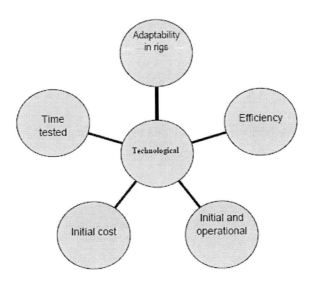

Figure 16. Technological indicators of sustainability

Pathway Analysis Product

Detailed pathway analysis of two technologies including their origin, degradation, oxidation and decomposition in order to demonstrate how the natural product is sustainable and the synthetic product is unsustainable. Two homologous products polyurethane fiber and wool fiber were selected for the sustainability assessment. They both appear to be the same in terms of durability, however, one is of natural origin and the other is made of hydrocarbon products synthetically. The pathways of these two products were studied including degradation behavior such as oxidation and photo degradation. A direct laboratory degradation experiment, and application of microwave on these products were also investigated.

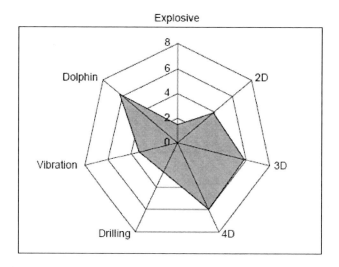

Figure 17. Proposed Spider web sustainability model to evaluated technology

OLYMPIC GREEN SUPPLY CHAIN

The supply chain of offshore oil and gas operations is developed on the basis of their lifecycle activities. Seismic exploration is at the top level of the chain and decommissioning is at the bottom most level. The intermediate steps include drilling, production, and transportation. The solid arrows on the left-hand side represent input, while the blank arrows on the right-hand side represent emission.

At present levels of operation, activities associated with the OOGOs supply chain emit numerous wastes into the air, water, and sediments. The seismic process emits 10 wastes, while the drilling process is the source of the highest emissions within the chain (Khan and Islam, 2006b). Major emissions from OOGOs, especially drilling wastes. The main wastes of OOGOs that impact the environment are drilling-waste fluids or muds, drilling-waste solids, produced water, and volatile organic compounds. Moreover, produced water constitutes the highest amount of waste released in the whole supply chain (Khan and Islam, 2005a and 2006b).

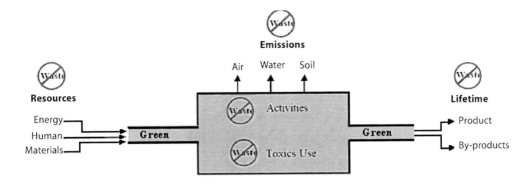

Figure 18. The proposed "Olympic" green supply chain for offshore oil and gas operations

Analogous to the "Olympic" logo, the proposed supply chain for OOGOs consists of five zeros (zero-wastes), which are: (i) zero emissions; (ii) zero resource wastes; (iii) zero wastes in activities; (iv) zero use of toxics; and (v) zero waste in product life-cycle. Figure 18 captures the concept of the Olympic supply chain for OOGOs in a representative diagram. According to this model, the system is enabled to input green resources. In Figure 18, energy, human, and materials are shown as green inputs. The rectangular box illustrated as a processing unit shows that toxic compounds cannot be used in the chain. Production and administrative activities can be wasted in regards to inefficiency in processing. The three arrows at the top of the picture indicate that the chain admits no gaseous, solid, and liquid emissions or discharges. The outlet of the system also shows green products that generate no product lifecycle, in the form of transportation, use, and end-of-use of the product. This supply chain model can be used to achieve sustainability in OOGOs. This is because such an approach is always the norm in nature. This is a new approach for use in the petroleum (Bjorndalen et al., 2005) and renewable energy sectors (Khan et al., 2005 and 2006).

CHALLENGES IN TECHNOLOGY DEVELOPMENT AND COMMERCIALIZATION

At present, it is a challenge to develop truly sustainable technology, because it is hard to break the present 'aphenomenal' technology development and marketing loop (Khan et al., 2005 and 2006). Recently developed technology development method gives a direction for developing a truly sustainable technology. However, there is a big gap between technology development and its marketing, which is pictorially shown in Figure 19. It is reported that a green technology can reach to consumer level due to present corporate driven aphnomenol marketing (Barkley research, ASME, 2005). If we are unable to overcome this gap, then a newly develop ground breaking sustainable technology can not see the daylight.

Moreover, this gap further widened due to the non-collaboration between technology inventor, and manufacturer. Generally, technologies are developed by academicians or researchers and manufactured by private or public sectors with the help of financial institution. Finally, that is consumed by community. To overcome the present problems all important components of technology development and marketing should be collaborated.

Figure 20 shows an integrated model, where academic, financial, private, community and public sector are integrated. The model will help to sustain a technology.

Figure 19. Gap between ground breaking sustainable technology development and its marketing

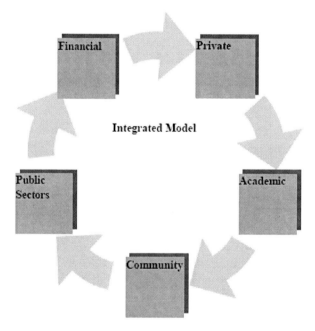

Figure 20. Collaboration among different sectors for sustainable technology development

New Supply Chain for Technology Development

At present, technologies are developed in a laboratory condition in an isolated manner. There is not collaboration among researcher, manufacture and consumer. However, to sustain a technology there should be collaboration well ahead from research to consumption stages. This integration can be achieved by using a technology supply chain. There is very few or hardly any work is done in the application of supply chain in the technology development. The main components of proposed supply chain are presented in Figure 21.

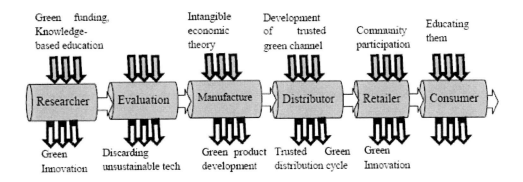

Figure 21. Technological supply chain

Research is the key element for developing innovative sustainable technology. The present trend of research for innovation does not take into account pro-nature approach in technology development. The reason behind this problem is that researcher is ill-educated. There is lacking of proper environmental, natural and science (not the technology) education in conventional academic curriculum. Even a university becomes a business corporate other than educational institute. As a result, the main goal of study is for earning money not seeking knowledge.

CONCLUSION

To achieve sustainability in technological development, a fair, consistent and scientifically acceptable criterion is needed. In this study, the time or temporal scale is considered as the prime selecting criterion for assuring inherent sustainability in technological development. The proposed model shows it to be feasible and that it could be easily applied to achieve true sustainability. This approach is particularly applicable to assess sustainable technology and other management tools, while the straightforward flowchart model being proposed should facilitate sustainability evaluation.

An evolutionary analysis of the historical record of the contemporary period suggests that technology and morality can and will respond to a clearly perceived future threat to civilization. However, we cannot easily predict the threat, or whether our response will be fast enough. Moving towards sustainable development requires that unpredictable/hidden environmental problems be resolved. For societies to attain or try to attain sustainable development, effort should be devoted to developing sustainable technology which is time tested. Development of this technology is easier, even though it is not generally encouraged in view of considering the initial cost of developing these types of technology. This concern regarding high cost of implementing truly sustainable technology is based on the currently used economic theories and business models that are themselves severely flowed and are indeed aphenomenal. As this paper will reveal, there are important developments that can produce a series of pro-nature and inherently sustainable technologies. But to ensure this eventuality, tough action motivated by an explicit moral concern for sustainability is required.

REFERENCES

Adrianto, L, Matsuda, Y and Sakuma, Y. 2005. Assessing local sustainability of .sherries system: a multi-criteria participatory approach with the case of Yoron Island, Kagoshima prefecture, Japan. *Marine Policy,* Vol. 29: 9-23.

Agarwal, P.N. and Puvathingal, J.M. 1969. Microbiological deterioration of woollen materials. *Textile Research Journal,* Vol. 39, 38.

Appleton, A.F., 2006. Sustainability: A practitioner's reflection, *Technology in Society*: in press.

Brown, L., Postel, S., and Flavin, C., 1991. From Growth To Sustainable Development. R. Goodland (editor), *Environmentally Sustainable Economic Development,* Paris: UNESCO, pp. 93-98.

Canesis, 2004. Wool the natural fiber, Canesis Network Ltd, Private Bag 4749, Christchurch, New Zealand.

Cariou, R., Jean-Philippe, A., Marchand, P., Berrebi, A., Zalko, D., Andre, F. and Bizec, B. 2005. New multiresidue analytical method dedicated to trace level measurement of brominated flame retardants in human biological matrices, *Journal of Chromatography*, *Vol.* 1100, No. 2: 144-152.

CDC (Centers for Disease Control) Report, 2001, National report on human exposure to environmental chemicals. Centers for Disease Control and Prevention, National Center for Environmental Health, Division of Laboratory Sciences, Mail Stop F-20, 4770 Buford Highway, NE, Atlanta, Georgia 30341-3724, NCEH Pub#: 05-0725.

CEIA (Centre d'Estudis d'Informaci Ambiental). 2001. A new model of environmental communication from Consumption to use of information. *European Environment Agency,* Copenhagen. 65 pp.

Chaalal, O., Tango, M., and Islam, M.R. 2005. A New Technique of Solar Bioremediation, *Energy Sources, vol.* 27, no. 3.

Chhetri, A.B., Khan, M.I., and Islam, M.R. 2006. A Novel Sustainably Developed Cooking Stove, *J. Nature Science and Sust. Technology*: submitted.

Costanza, R., J. Cumberland, H. Daly, R. Goodland and R. Norgaard, 1997. An introduction to ecological economics, international society for ecological economics and St. Lucie Press, Florida.

CWRT (Center for Waste Reduction Technologies, AIChE). 2002. Collaborative Projects: Focus area Sustainable Development: *Development of Baseline Metrics;* 2002. <http://www.aiche.org/cwrt/pdf/ BaselineMetrics.pdf> [Accessed: May 28, 2005]

Daly, H. E. 1992. Allocation. distribution. and scale: towards an economics that is efficient. just and sustainable. *Ecological Economics,* Vol. 6: 185-193.

Daly, H.E., 1999. Ecological economics and the ecology of cconomics, essay in criticism, Edward Elgar, U.K.

Darton R., 2002. Sustainable development and energy: predicting the future. In. Proceedings of the 15th International Conference of Chemical and Process Engineering; 2002.

Dewulf J, Van Langenhove H, Mulder J, van den Berg MMD, van der Kooi HJ, de Swaan Arons J., 2002. *Green Chem, Vol.* 2:108–14.

Dewulf, J. and Langenhove, H. V. 2004. Integrated industrial ecology principles into a set of environmental sustainability indicators for technology assessment. *Resource Conservation and Recycling:* in press.

Donnelly, K., Beckett-Furnell, Z., Traeger, S., Okrasinski T. and Holman, S., 2006. Eco-design implemented through a product-based environmental management system, *Journal of Cleaner Production,* Vol xx: 1-11.

Eissen M, Metzger J.O., 2002. Environmental performance metrics for daily use in synthetic chemistry. *Chem. Eur. J.* 2002;8:3581–5.

Gibson, R. 1991. Should Environmentalists Pursue Sustainable Development? *Probe Post,* 22-25.

Gleick, J., 1987, Chaos – making a new science, Penguin Books, NY, 352 pp.

GRI (Global Reporting Initiative). 2002. Sustainability Reporting Guidelines. GRI: Boston.

Hawken, P., 1992. The Ecology of Commerce. New York: Plenum.

IChemE (Institute of Chemical Engineers). 2002. The sustainability metrics, sustainable development progress metrics recommended for use in the process industries; 2002. <http://www.getf.org/file/toolmanager/O16F26202.pdf.> [Accessed: June 10,2005]

Islam, M.R. and Farouq Ali, S.M., 1991, Scaling of in-situ combustion experiments, *J. Pet. Sci.Eng.*, vol. 6, 367-379.

Islam, M.R., 2003. Revolution in Education, Halifax, Canada, EECRG Publication, Nova Scotia, Canada (ISBN 0-9733656-1-7), 553 pp..

Islam, M.R., 2004, Inherently-sustainable energy production schemes, *EEC Innovation,* vol. 2, no. 3, 38-47.

Islam, M.R., 2005a. Knowledge-based technologies for the information age, JICEC05-Keynote speech, Jordan International Chemical Engineering Conference V, 12-14 September 2005, Amman, Jordan.

Islam, M.R., 2005b, Unraveling the mysteries of chaos and change: knowledge-based technology development", *EEC Innovation,* vol. 2, no. 2 and 3, 45-87.

Islam, M.R., 2005c. Developing knowledge-based technologies for the information age through virtual universities, E-transformation Conference, April 21, 2005. Turkey.

Islam, M.R., 2005d. Unravelling the mysteries of chaos and change: the knowledge-based technology development, Proceedings of the First International Conference on Modeling, Simulation and Applied Optimization, Sharjah, U.A.E. February 1-3, 2005.

Islam, M.R., Verma. A., and Farouq Ali, S.M., 1991, In situ combustion - the essential reaction kinetics", Heavy Crude and Tar Sands - Hydrocarbons for the 21st Century, vol. 4, UNITAR/UNDP.

Judes, U., 2000. Towards a culture of sustainability. W. L. Leal Filho (editor), *Communicating Sustainability,* Vol. 8pp. 97-121. Berlin: Peter Lang.

Khan, M.I. and Islam, M.R,. 2005a. Assessing sustainability of technological developments: an alternative approach of selecting indicators in the case of Offshore operations. ASME Congress, 2005, Orlando, Florida, Nov 5-11, 2005, Paper no.: IMECE2005-82999.

Khan, M.I. and Islam, M.R., 2005b. Sustainable marine resources management: framework for environmental sustainability in offshore oil and gas operations. Fifth International Conference on Ecosystems and Sustainable Development. Cadiz, Spain, May 03-05, 2005.

Khan, M.I. and Islam, M.R., 2005c. Achieving True technological sustainability: pathway analysis of a sustainable and an unsustainable product, International Congress of Chemistry and Environment, Indore, India, 24-26 December 2005. (Acc. Abs. No. 98).

Khan, M.I, Zatzman, G. and Islam, M.R., 2005. New sustainability criterion: development of single sustainability criterion as applied in developing technologies. Jordan International Chemical Engineering Conference V, Paper No.: JICEC05-BMC-3-12, Amman, Jordan, 12 - 14 September 2005.

Khan, M.I. and Islam, M.R., 2006a. Developing sustainable technologies for offshore seismic operations, *Journal of Petroleum Science and Technology*: accepted for publication, December, 22 pp.

Khan, M.I. and Islam, M.R., 2006b. *True Sustainability in Technological Development and Natural Resources Management*. Nova Science Publishers, New York: 355 pp. [ISBN: 1-60021-203-4].

Khan, M.I. and Islam, M.R., 2007. *Handbook of Sustainable Oil and Gas Operations Management*, Gulf Publishing Company, Austin, Texas: in press. [Scheduled to be published in April, 2007].

Khan, M.I., Chhetri, A.B., and Islam, M.R., 2006a. Analyzing sustainability of community-based energy development technologies, *Energy Sources*: in press.

Khan, M.I., Lakhal, Y.S., Satish, M., and Islam, M.R., 2006b. Towards achieving sustainability: application of green supply chain model in offshore oil and gas operations. *Int. J. Risk Assessment and Management:* in press.

Kunisue, T., Masayoshi Muraoka, Masako Ohtake, Agus Sudaryanto, Nguyen Hung Minh, Daisuke Ueno, Yumi Higaki, Miyuki Ochi, Oyuna Tsydenova, Satoko Kamikawa et al., 2006. Contamination status of persistent organochlorines in human breast milk from Japan: Recent levels and temporal trend, *Chemosphere*: in press.

Labuschange, C. Brent, A.C. and Erck, R.P.G., 2005. Assessing the sustainability performances of industries. *Journal of Cleaner Production*, Vol. 13: 373-385.

Lakhal, S., S. H'mida and R. Islam, 2005. A Green supply chain for a petroleum company, Proceedings of 35th International Conference on Computer and Industrial Engineering, Istanbul, Turkey, June 19-22, 2005, Vol. 2: 1273-1280.

Lange J-P. Sustainable development: efficiency and recycling in chemicals manufacturing. *Green Chem,* 2002; 4:546–50.

Leal Filho, W. L. (1999). Sustainability and university life: some European perspectives. W. Leal Filho (ed.), Sustainability and University Life: Environmental Education, Communication and Sustainability (pp. 9-11). Berlin: Peter Lang.

Lems S, van derKooi HJ, deSwaan Arons J. 2002. The sustainability of resource utilization. *Green Chem,* Vol.4: 308–13.

Leontieff, W. 1973. Structure of the world economy: outline of a simple input-output formulation, Stockholm: Nobel Memorial Lecture, 11 December, 1973.

Livingston, R.J., and Islam, M.R., 1999, Laboratory modeling, field study and numerical simulation of bioremediation of petroleum contaminants, *Energy Sources*, vol. 21 (1/2), 113-130.

Lowe EA, Warren JL, Moran SR. Discovering industrial ecology—an executive briefing and sourcebook. Columbus: Battelle Press; 1997.

Lowy, J. 2004. Plastic left holding the bag as environmental plague. Nations around world look at a ban. <http://seattlepi.nwsource.com/national/182949_bags21.html>.

Lubchenco, J. A., et al. 1991. The sustainable biosphere initiative: an ecological research agenda. *Ecology* 72:371- 412.

Lunder, S. and Sharp, R., 2003. Mother's milk, record levels of toxic fire retardants found in American mother's breast milk. Environmental Working Group, Washington, USA.

Market Development Plan, 1996. Market status report: postconsumer plastics, business waste reduction, Integrated Waste Development Board, Public Affairs Office. California.

Marx, K. 1883. Capital: A critique of political economy Vol. II: The Process of Circulation of Capital, (London, Edited by Frederick Engels.

Matsuoka, K., Iriyama, Y., Abe, T., Matsuoka, M., Ogumi, Z., 2005. Electro-oxidation of methanol and ethylene glycol on platinum in alkaline solution: Poisoning effects and product analysis. *Electrochimica Acta,* Vol.51: 1085–1090.

McCarthy, B.J., Greaves, P.H., 1988. Mildew-causes, detection methods and prevention. *Wool Science Review,* Vol. 85, 27–48.

MEA (Millennium Ecosystem Assessment), 2005. The millennium ecosystem assessment, Commissioned by the United Nations, the work is a four-year effort by 1,300 scientists from 95 countries.

Miller, G. (1994). Living in the Environment: Principles, Connections and Solutions. California: Wadsworth Publishing.

Mittelstaedt, M., 2006. Chemical used in water bottles linked to prostate cancer, *The Globe and Mail,* Friday, 09 June 2006.

Molero, C., Lucas, A. D. and Rodrıguez, J. F., 2006. Recovery of polyols from flexible polyurethane foam by "split-phase" glycolysis: Glycol influence. *Polymer Degradation and Stability.* Vol. 91: 221-228.

Narayan, R., 2004. Drivers and rationale for use of biobased materials based on life cycle asessment (LCA). GPEC 2004 Paper.

Nikiforuk, A. (1990). Sustainable Rhetoric. Harrowsmith, 14-16.

OCED, 1998. Towards sustainable development: environmental indicators. Paris: Organization for Economic Cooperation and Development; 132pp.

OECD, 1993. Organization for Economic Cooperation and development core set of indicators for environmental performance reviews. A synthesis report by the Group on State of the Environment,. Paris, 1993.

Plastic Task Force (1999). Adverse health effects of plastics. <http://www.ecologycenter.org/erc /fact_sheets plastichealtheffects.html# plastichealthgrid>

Pokharel, G.R., Chhetri, A, B., Devkota, S. and Shrestha, P., 2003. En route to strong sustainability: can decentralized community owned micro hydro energy systems in Nepal Realize the Paradigm? A case study of Thampalkot VDC in Sindhupalchowk District in Nepal. International Conference on Renewable Energy Technology for Rural Development. Kathmandu, Nepal.

Pokharel, G.R., Chhetri, A.B., Khan, M.I., and Islam, M.R., 2006. Decentralized micro hydro energy systems in Nepal: en route to sustainable energy development, *Energy Sources - Part B: Economics, Planning and Policy*: accepted for publication, March, 18 pp.

Rahbur S., Khan, M.M., M. Satish, Ma, F. and Islam, M.R., 2005. Experimental and numerical studies on natural insulation materials, ASME Congress, 2005, Orlando, Florida, Nov 5-11, 2005. *IMECE* 2005-82409.

Rees, W. (1989). Sustainable development: myths and realities. Proceedings of the Conference on Sustainable Development Winnipeg, Manitoba: IISD.

Robinson, J. G. 1993. The limits to caring: sustainable living and the loss of biodiversity. *Conservation Biology* 7: 20- 28.

Saito, K, Ogawa, M. Takekuma, M., Ohmura, A., Migaku Kawaguchi a, Rie Ito a, Koichi Inoue a, Yasuhiko Matsuki c, Hiroyuki Nakazawa, 2005. Systematic analysis and overall toxicity evaluation of dioxins and hexachlorobenzene in human milk, *Chemosphere,* Vol. 61: 1215–1220.

Smith P. How green is my process? A practical guide to green metrics. In: Proceedings of the Conference Green Chemistry on Sustainable Products and Processes; 2001.

Spangenberg, J.H. and Bonniot, O., 1998. Sustainability indicators-a compass on the road towards sustainability. Wuppertal Paper No. 81, February 1998. *ISSN* No. 0949-5266.

Sraffa, P. 1960. Production of Commodities by Means of Commodities. Cambridge, Cambridge University Press.

Sudaryanto, A., Tatsuya Kunisue, Natsuko Kajiwara, Hisato Iwata, Tussy A. Adibroto, Phillipus Hartono and Shinsuke Tanabe, 2006. Specific accumulation of organochlorines in human breast milk from Indonesia: Levels, distribution, accumulation kinetics and infant health risk, *Environmental Pollution,* Vol. 139, No. 1: 107-117

Szostak-Kotowa, J., 2004. Biodeterioration of textiles, International Biodeterioration and Biodegradation, Vol. 53: 165 – 170.

UNCSD (United Nations Commission on Sustainable Development), 2001. Indicators of Sustainable Development: Guidelines and Methodologies, United Nations, New York.

Wackernagel, M., and Rees, W. (1996). Our ecological footprint. Gabriola Island: New Society Publishers.

Waste Online, 2005. Plastic recycling information sheet. < http://www.wasteonline. org.uk/ resources/InformationSheets/Plastics.htm> [Accessed: February 20, 2006].

WCED (World Commission on Environment and Development) 1987. Our common future. World Conference on Environment and Development. Oxford: Oxford University Press; 1987. 400pp.

Welford, R. (1995). Environmental strategy and sustainable development: the corporate challenge for the 21st Century. London: Routledge.

Winterton N., 2001. Twelve more green chemistry principles. *Green Chem. Vol.* 3: G73–5.

World Health Organization (WHO). 1994. Brominated diphenyl ethers. Environmental Health Criteria, Vol.162. International Program on Chemical Safety.

Wright, T. 2002. Definitions and frameworks for environmental sustainability in Higher education. *International Journal of Sustainability in Higher Education Policy*, Vol. 15 (2).

Zatzman, G.M. and Islam, M.R., 2006, *Economics of Intangibles*, Nova Science Publishers, New York: in press.

In: Nature Science and Sustainable Technology
Editor: M. R. Islam, pp. 39-71

ISBN: 978-1-60456-009-1
© 2008 Nova Science Publishers, Inc.

Chapter 3

THE ENGINEERING APPROACH VERSUS THE MATHEMATICAL APPROACH IN DEVELOPING RESERVOIR SIMULATORS

*J. H. Abou-Kassem**

Chemical and Petroleum Engineering Department, UAE University,
Al-Ain, P.O. Box 17555, United Arab Emirates

ABSTRACT

Reservoir simulation in the oil industry has become the standard for solving reservoir- engineering problems. Reservoir simulation combines physics, mathematics, reservoir engineering, and computer programming to develop a tool for predicting hydrocarbon reservoir performance under various production strategies. The steps involved in the development of a simulator include: derivation of the partial differential equations (PDE's) describing the recovery process through formulation, discretization of the PDE's in space and time to obtain nonlinear algebraic equations, linearization of resulting algebraic equations, solving the linearized algebraic equations numerically, and finally validation of the simulator. Developers of simulators relied heavily on mathematics in the first two steps (mathematical approach) to obtain the third step (nonlinear algebraic equations or finite-difference equations). A new approach, that derives the finite-difference equations without going through the rigor of PDE's and discretization, is presented in this paper. The new approach is called the engineering approach because it is closer to the engineer's thinking and to the physical meaning of the equations. Both the engineering and mathematical approaches treat boundary conditions with the same accuracy if the mathematical approach uses second order approximations. The engineering approach is simple and yet general and rigorous. In addition, it results in the same finite-difference equations for any hydrocarbon recovery process. Because the engineering approach is independent of the mathematical approach, it provides justification for the use of central differencing in space, and gives implications of the

---

[*] Corresponding author: Email: J.Aboukassem@uaeu.ac.ae

approximations, that are usually used in the mathematical approach, in time discretization.

INTRODUCTION

Reservoir Simulation is an art of combining physics, mathematics, reservoir engineering, and computer programming to develop a tool for predicting hydrocarbon reservoir performance under various operating conditions. Fig. 1 depicts the major steps involved in the development of a reservoir simulator (Odeh, 1982). In this figure, the *formulation* step outlines the basic assumptions in the system and applies them to a control volume in a heterogeneous reservoir. The result of this step is a set of coupled, nonlinear partial differential equations (PDE's) that describes fluid flow through porous media.

The PDE's derived during the formulation step, if solved analytically, would give the reservoir pressure, fluid saturations, and well production rates as continuous functions of space and time. Because of the nonlinear nature of the PDE's, analytical techniques cannot be used and solutions must be obtained with numerical methods. In contrast to analytical solutions, numerical solutions give the values of pressure and fluid saturations only at discrete points in the reservoir at discrete times. *Discretization* is the process of converting the PDE's into algebraic equations. Several numerical methods can be used to discretize the PDE's; however, the most common approach in the oil industry today is the finite-difference method. The discretization process results in a system of nonlinear algebraic equations. These equations generally cannot be solved with algebraic techniques and linearization of such equations becomes a necessary step before solutions can be obtained. *Linearization* involves approximating nonlinear terms (transmissibilities and coefficients of unknowns in the accumulation term) in both space and time. Once the nonlinear algebraic equations have been linearized, any one of several linear equation solvers can be used to obtain their *solution*. *Validation* of a reservoir simulator is the last step in developing a simulator, after which the simulator can be used for practical field applications. The validation step is necessary to make sure that no errors were introduced in the various aforementioned steps and in computer programming.

Basically, there are three methods available for the discretization of any PDE: the Taylor series method, the integral method, and the variational method (Aziz and Settari, 1979). The first two methods result in the finite-difference method, whereas the third method results in the variational method. The discretization converts nonlinear PDE's into nonlinear algebraic equations. In this paper, the term "Mathematical Approach" refers to the methods that obtain the nonlinear algebraic equations through deriving and discretizing the PDE's. The "Engineering Approach" presented in this paper arrives at the same nonlinear algebraic equations without the need to resort to PDE's or their discretization. The treatment of boundary conditions by the engineering approach will be presented and compared with that by the mathematical approach.

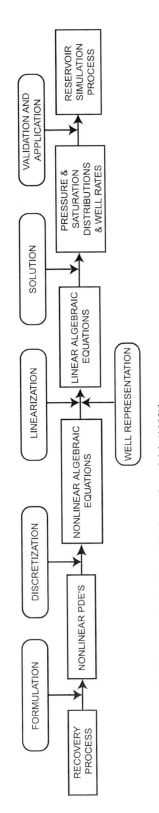

Figure 1. Major steps used to develop reservoir simulators. [Redrawn from Odeh (1982)]

DERIVATION OF FLUID FLOW EQUATIONS IN DISCRETIZED FORM

The fluid flow equations in discretized form (nonlinear algebraic equations) can be obtained by either the traditional mathematical approach or the proposed engineering approach. Both of these approaches make use of the same basic principles and both approaches discretize the reservoir into gridblocks (or gridpoints). Both approaches yield the same discretized flow equations for modeling any reservoir-fluid system (multiphase, multi component, thermal, heterogeneous reservoir) using any coordinate system (Cartesian, cylindrical, spherical) in one-dimensional (1D), two-dimensional (2D), or three-dimensional (3D) reservoirs (Abou-Kassem, Farouq Ali, and Islam, 2006). Therefore, the presentation here will be for modeling the flow of single-phase, compressible fluid in horizontal, 1D reservoir using irregular block size distribution in rectangular coordinates. We will take advantage of this simple case to demonstrate the capacity of the engineering approach to give justification for and interpretations of the discretization methods used in the mathematical approach, and to present treatment of boundary conditions.

Basic Principles

The basic principles include mass conservation, equation of state, and constitutive equation. The principle of mass conservation states that the total mass of fluid entering and leaving a volume element of the reservoir must equal the net increase in the mass of the fluid in the reservoir element,

$$m_i - m_o + m_s = m_a, \qquad (1)$$

where m_i = "mass in" = the mass of fluid entering the reservoir element from other parts of the reservoir, m_o = "mass out" = the mass of fluid leaving the reservoir element to other parts of the reservoir, m_s = "sink/source" = the mass of fluid entering or leaving the reservoir element externally through wells, and m_a = "mass accumulated" = the mass of excess fluid stored in or depleted from the reservoir element over a time interval.

An equation of state describes the density of fluid as a function of pressure and temperature,

$$B = \rho_{sc} / \rho, \qquad (2)$$

where B = fluid formation volume factor which is a function of pressure [B = f (p)], ρ = fluid density at reservoir conditions, and ρ_{sc} = fluid density at standard conditions.

A constitutive equation describes the rate of fluid movement into (or out of) the reservoir element. In reservoir simulation, Darcy's law is used to relate fluid flow rate to potential gradient. The differential form of Darcy's law for a horizontal reservoir is

$$u_x = q_x / A_x = -\beta_c \frac{k_x}{\mu} \frac{\partial p}{\partial x}, \qquad (3)$$

where β_c = unit conversion factor for the transmissibility coefficient, k_x = absolute permeability of rock along the direction of flow, μ = fluid viscosity, p = pressure, and u_x = superficial velocity of fluid defined as fluid flow rate (q_x) per unit cross-sectional area (A_x) normal to flow direction x.

Reservoir Discretization

Reservoir discretization means that the reservoir is described by a set of gridblocks (or gridpoints) whose properties, dimensions, boundaries, and locations in the reservoir are well defined. Fig. 2 shows reservoir discretization in the x-direction for both block-centered and point-distributed grids in rectangular coordinates as one focuses on Gridblock i or Gridpoint i. The figure shows how the blocks are related to each others [Block i and its neighboring blocks (Blocks $i-1$ and $i+1$)], block dimensions (Δx_i, Δx_{i-1}, Δx_{i+1}), block boundaries ($x_{i-1/2}$, $x_{i+1/2}$), distances between the point that represents the block and block boundaries (δx_{i^-}, δx_{i^+}), and distances between the gridpoints or points representing the blocks ($\Delta x_{i-1/2}$, $\Delta x_{i+1/2}$). In addition, each gridblock or gridpoint is assigned elevation and rock properties such as porosity and permeabilities.

In block-centered grid system, the grid is constructed by choosing n_x gridblocks that span the entire reservoir length in the x-direction. The gridblocks are assigned predetermined dimensions (Δx_i, $i = 1, 2, 3... n_x$) that are not necessarily equal. Then the point that represents a gridblock is consequently located at the center of the gridblock. In point-distributed grid system, the grid is constructed by choosing n_x gridpoints that span the entire reservoir length in the x-direction. In other words, the first gridpoint is placed at one reservoir boundary and the last gridpoint is placed at the other reservoir boundary. The distances between gridpoints are assigned predetermined values ($\Delta x_{i+1/2}$, $i = 1, 2, 3... n_x-1$) that are not necessarily equal. Each gridpoint represents a gridblock whose boundaries are placed halfway between the gridpoint and its neighboring gridpoints.

THE CURRENT APPROACH (MATHEMATICAL APPROACH)

In the mathematical approach, the algebraic flow equations are derived in three consecutive steps: 1) derivation of the PDE describing fluid flow in reservoir using the three basic principles mentioned earlier, 2) discretization of reservoir into gridblocks or gridpoints as shown above, and 3) discretization of the resulting PDE in space and time.

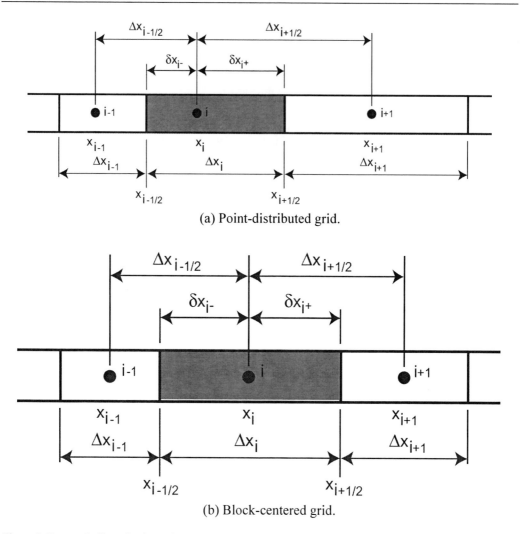

Figure 2. Reservoir discretization using: (a) point-distributed grid, (b) block-centered grid.

Derivation of PDE

Fig. 3 shows a finite control volume with a cross-sectional area A_x perpendicular to the direction of flow, length Δx in the direction of flow, and volume $V_b = A_x \Delta x$. Point x represents the left side of the control volume. The fluid enters the control volume across its surface at x and leaves across its surface at $x + \Delta x$ at mass rates of $w_x|_x$ and $w_x|_{x+\Delta x}$, respectively. The fluid also enters the control volume through a well at a mass rate of q_m. The mass of fluid in the control volume per unit volume of rock is m_v. Therefore, the material balance equation written over a time step Δt as expressed by Eq. 1 becomes

$$m_i|_x - m_o|_{x+\Delta x} + m_s = m_a \qquad (4)$$

or

$$w_x\big|_x \Delta t - w_x\big|_{x+\Delta x}\Delta t + q_m \Delta t = m_a, \qquad (5)$$

where mass flow rate (w_x) and mass flux ($\dot{m}_x$) are related through

$$w_x = \dot{m}_x A_x. \qquad (6)$$

In addition, mass accumulation is defined as

$$m_a = \Delta_t(V_b m_v) = V_b(m_v\big|_{t+\Delta t} - m_v\big|_t) = V_b(m_v^{n+1} - m_v^n). \qquad (7)$$

Substitution of Eqs. 6 and 7 into Eq. 5 yields

$$(\dot{m}_x A_x)\big|_x \Delta t - (\dot{m}_x A_x)\big|_{x+\Delta x}\Delta t + q_m \Delta t = V_b(m_v\big|_{t+\Delta t} - m_v\big|_t). \qquad (8)$$

Dividing Eq. 8 by $V_b \Delta t$, observing that $V_b = A_x \Delta x$, and rearranging results in

$$-[(\dot{m}_x\big|_{x+\Delta x} - \dot{m}_x\big|_x)/\Delta x] + \frac{q_m}{V_b} = [(m_v\big|_{t+\Delta t} - m_v\big|_t)/\Delta t]. \qquad (9)$$

The limits of the terms in brackets in above equation as Δx and Δt approach zero (i.e., as $\Delta x \to 0$ and $\Delta t \to 0$) become first order partial derivatives and the resulting equation becomes

$$-\frac{\partial \dot{m}_x}{\partial x} + \frac{q_m}{V_b} = \frac{\partial m_v}{\partial t}. \qquad (10)$$

Mass flux ($\dot{m}_x$) can be stated in terms of fluid density (ρ) and volumetric velocity (u_x) as

$$\dot{m}_x = \alpha_c \rho u_x, \qquad (11)$$

m_v can be expressed in terms of fluid density and porosity (ϕ) as

$$m_v = \phi \rho, \qquad (12)$$

and q_m can be expressed in terms of well volumetric rate (q) and fluid density as

$$q_m = \alpha_c \rho q. \tag{13}$$

Using Eqs. 11 through 13, Eq. 10 can be rewritten in another form known as the continuity equation,

$$-\frac{\partial(\rho u_x)}{\partial x} + \frac{\rho q}{V_b} = \frac{1}{\alpha_c}\frac{\partial(\rho \phi)}{\partial t}. \tag{14}$$

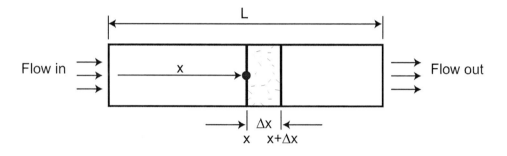

Figure 3. Control volume in 1D traditionally used for writing material balance. [Redrawn from Aziz and Settari (1979)]

The flow equation can be obtained by combining the continuity equation (Eq. 14), the equation of state (Eq. 2), and Darcy's law (Eq. 3), and noting that $q/B = q_{sc}$. The resulting flow equation for single phase flow is

$$\frac{\partial}{\partial x}(\beta_c \frac{k_x}{\mu B}\frac{\partial p}{\partial x}) + \frac{q_{sc}}{V_b} = \frac{1}{\alpha_c}\frac{\partial}{\partial t}(\frac{\phi}{B}). \tag{15}$$

Above equation is the PDE that describes single-phase flow in 1D rectangular coordinates.

Discretization of PDE in Space and Time

First, the reservoir is discretized as mentioned earlier. Second, Eq. 15 is rewritten in another form, to take care of variations of cross-sectional area through multiplying by $V_b = A_x \Delta x$, as

$$\frac{\partial}{\partial x}(\beta_c \frac{k_x A_x}{\mu B}\frac{\partial p}{\partial x})\Delta x + q_{sc} = \frac{V_b}{\alpha_c}\frac{\partial}{\partial t}(\frac{\phi}{B}) \tag{16}$$

Eq. 16 is then written for Gridblock i,

$$\frac{\partial}{\partial x}(\beta_c \frac{k_x A_x}{\mu B}\frac{\partial p}{\partial x})_i \Delta x_i + q_{sc_i} = \frac{V_{b_i}}{\alpha_c}\frac{\partial}{\partial t}(\frac{\phi}{B})_i. \tag{17}$$

Space Discretization

The second order derivative w.r.t. x at Point i appearing on the LHS of Eq. 17 is approximated using second order central differencing. The resulting approximation can be written as

$$\frac{\partial}{\partial x}(\beta_c \frac{k_x A_x}{\mu B}\frac{\partial p}{\partial x})_i \Delta x_i \cong T_{x_{i-1/2}}(p_{i-1}-p_i) + T_{x_{i+1/2}}(p_{i+1}-p_i) \tag{18}$$

with transmissibility $T_{x_{i\mp1/2}}$ being defined as

$$T_{x_{i\mp1/2}} = (\beta_c \frac{k_x A_x}{\mu B \Delta x})_{i\mp1/2}. \tag{19}$$

The process of the approximation leading to Eq. 18 can be looked at as follows. Using the definition of central-difference approximation to the first order derivative evaluated at Point i (see Fig. 2), one can write

$$\frac{\partial}{\partial x}(\beta_c \frac{k_x A_x}{\mu B}\frac{\partial p}{\partial x})_i \cong [(\beta_c \frac{k_x A_x}{\mu B}\frac{\partial p}{\partial x})_{i+1/2} - (\beta_c \frac{k_x A_x}{\mu B}\frac{\partial p}{\partial x})_{i-1/2}]/\Delta x_i. \tag{20}$$

Use of central differencing again to approximate $(\frac{\partial p}{\partial x})_{i\mp1/2}$ yields

$$(\frac{\partial p}{\partial x})_{i+1/2} \cong (p_{i+1}-p_i)/(x_{i+1}-x_i) = (p_{i+1}-p_i)/\Delta x_{i+1/2} \tag{21}$$

and

$$(\frac{\partial p}{\partial x})_{i-1/2} \cong (p_i - p_{i-1})/(x_i - x_{i-1}) = (p_i - p_{i-1})/\Delta x_{i-1/2}. \tag{22}$$

Substitution of Eqs. 21 and 22 into Eq. 20 and rearranging results in

$$\frac{\partial}{\partial x}(\beta_c \frac{k_x A_x}{\mu B} \frac{\partial p}{\partial x})_i \Delta x_i \cong [(\beta_c \frac{k_x A_x}{\mu B \Delta x})_{i+1/2}(p_{i+1} - p_i) - (\beta_c \frac{k_x A_x}{\mu B \Delta x})_{i-1/2}(p_i - p_{i-1})] \qquad (23)$$

or

$$\frac{\partial}{\partial x}(\beta_c \frac{k_x A_x}{\mu B} \frac{\partial p}{\partial x})_i \Delta x_i \cong [(\beta_c \frac{k_x A_x}{\mu B \Delta x})_{i+1/2}(p_{i+1} - p_i) + (\beta_c \frac{k_x A_x}{\mu B \Delta x})_{i-1/2}(p_{i-1} - p_i)]. \qquad (24)$$

Eq. 18 results from the substitution of $T_{x_{i\mp 1/2}}$ given by Eq. 19 into Eq. 24.

Substitution of Eq. 18 into the PDE given by Eq. 17 yields an equation that is discrete in space but continuous in time,

$$T_{x_{i-1/2}}(p_{i-1} - p_i) + T_{x_{i+1/2}}(p_{i+1} - p_i) + q_{sc_i} \cong \frac{V_{b_i}}{\alpha_c} \frac{\partial}{\partial t}(\frac{\phi}{B})_i. \qquad (25)$$

Time Discretization

The discretization of Eq. 25 in time is accomplished by approximating the first order derivative appearing on the RHS of the equation. We will consider here the forward-difference, backward-difference, and central-difference approximations. All three approximations can be written as

$$\frac{\partial}{\partial t}(\frac{\phi}{B})_i \cong \frac{1}{\Delta t}[(\frac{\phi}{B})_i^{n+1} - (\frac{\phi}{B})_i^n]. \qquad (26)$$

Forward-Difference Discretization

In the forward-difference discretization, one writes Eq. 25 at Time Level n (old time level t^n),

$$[T_{x_{i-1/2}}(p_{i-1} - p_i) + T_{x_{i+1/2}}(p_{i+1} - p_i) + q_{sc_i}]^n \cong \frac{V_{b_i}}{\alpha_c}[\frac{\partial}{\partial t}(\frac{\phi}{B})_i]^n. \qquad (27)$$

In this case, it can be looked at Eq. 26 as forward-difference of the first order derivative w.r.t. time at Time Level n. The discretized flow equation is called a forward-central-difference equation,

$$T_{x_{i-1/2}}^n (p_{i-1}^n - p_i^n) + T_{x_{i+1/2}}^n (p_{i+1}^n - p_i^n) + q_{sc_i}^n \cong \frac{V_{b_i}}{\alpha_c \Delta t}[(\frac{\phi}{B})_i^{n+1} - (\frac{\phi}{B})_i^n]. \qquad (28)$$

The RHS of Eq. 28 can be expressed in terms of the pressure of Gridblock i such that material balance is preserved. The resulting equation is

$$T^n_{x_{i-1/2}}(p^n_{i-1}-p^n_i)+T^n_{x_{i+1/2}}(p^n_{i+1}-p^n_i)+q^n_{sc_i} \cong \frac{V_{b_i}}{\alpha_c \Delta t}(\frac{\phi}{B})'_i[p^{n+1}_i-p^n_i], \quad (29)$$

where the derivative $(\frac{\phi}{B})'_i$ is defined as the chord; i.e.,

$$(\frac{\phi}{B})'_i = [(\frac{\phi}{B})^{n+1}_i - (\frac{\phi}{B})^n_i]/[p^{n+1}_i - p^n_i]. \quad (30)$$

Backward-Difference Discretization

In the backward-difference discretization, one writes Eq. 25 at Time Level $n+1$ (current time level t^{n+1}),

$$[T_{x_{i-1/2}}(p_{i-1}-p_i)+T_{x_{i+1/2}}(p_{i+1}-p_i)+q_{sc\,i}]^{n+1} \cong \frac{V_{b_i}}{\alpha_c}[\frac{\partial}{\partial t}(\frac{\phi}{B})_i]^{n+1}. \quad (31)$$

In this case, it can be looked at Eq. 26 as backward-difference of the first order derivative w.r.t. time at Time Level $n+1$. The discretized flow equation is called a backward-central-difference equation,

$$T^{n+1}_{x_{i-1/2}}(p^{n+1}_{i-1}-p^{n+1}_i)+T^{n+1}_{x_{i+1/2}}(p^{n+1}_{i+1}-p^{n+1}_i)+q^{n+1}_{sc_i} \cong \frac{V_{b_i}}{\alpha_c \Delta t}[(\frac{\phi}{B})^{n+1}_i - (\frac{\phi}{B})^n_i]. \quad (32)$$

The equation that corresponds to Eq. 29 is

$$T^{n+1}_{x_{i-1/2}}(p^{n+1}_{i-1}-p^{n+1}_i)+T^{n+1}_{x_{i+1/2}}(p^{n+1}_{i+1}-p^{n+1}_i)+q^{n+1}_{sc_i} \cong \frac{V_{b_i}}{\alpha_c \Delta t}(\frac{\phi}{B})'_i[p^{n+1}_i-p^n_i]. \quad (33)$$

Central-Difference Discretization

In the central-difference discretization, one writes Eq. 25 at the Time Level $n+1/2$ (time level $t^{n+1/2}$),

$$[T_{x_{i-1/2}}(p_{i-1}-p_i)+T_{x_{i+1/2}}(p_{i+1}-p_i)+q_{sc\,i}]^{n+1/2} \cong \frac{V_{b_i}}{\alpha_c}[\frac{\partial}{\partial t}(\frac{\phi}{B})_i]^{n+1/2}. \quad (34)$$

In this case, it can be looked at Eq. 26 as central-difference of the first order derivative w.r.t. time at Time Level $n+1/2$. In addition, the flow terms at Time Level $n+1/2$ are approximated by the average values at Time Level $n+1$ and Time Level n. The discretized flow equation in this case is the Crank-Nicholson approximation,

$$(1/2)[T^n_{x_{i-1/2}}(p^n_{i-1}-p^n_i)+T^n_{x_{i+1/2}}(p^n_{i+1}-p^n_i)]+(1/2)[T^{n+1}_{x_{i-1/2}}(p^{n+1}_{i-1}-p^{n+1}_i)+T^{n+1}_{x_{i+1/2}}(p^{n+1}_{i+1}-p^{n+1}_i)]$$
$$+(1/2)[q^n_{sc_i}+q^{n+1}_{sc_i}] \cong \frac{V_{b_i}}{\alpha_c \Delta t}[(\frac{\phi}{B})^{n+1}_i - (\frac{\phi}{B})^n_i]. \tag{35}$$

The equation that corresponds to Eq. 29 is

$$(1/2)[T^n_{x_{i-1/2}}(p^n_{i-1}-p^n_i)+T^n_{x_{i+1/2}}(p^n_{i+1}-p^n_i)]+(1/2)[T^{n+1}_{x_{i-1/2}}(p^{n+1}_{i-1}-p^{n+1}_i)+T^{n+1}_{x_{i+1/2}}(p^{n+1}_{i+1}-p^{n+1}_i)]$$
$$+(1/2)[q^n_{sc_i}+q^{n+1}_{sc_i}] \cong \frac{V_{b_i}}{\alpha_c \Delta t}(\frac{\phi}{B})'_i[p^{n+1}_i - p^n_i]. \tag{36}$$

Observations on the Derivation of the Mathematical Approach

1. For a discretized reservoir having heterogeneous block permeability distribution and irregular grid blocks (neither constant nor equal Δx), blocks have defined dimensions and permeabilities; therefore, interblock geometric factor $[(\beta_c \frac{k_x A_x}{\Delta x})|_{x_{i\mp1/2}}]$ is constant, independent of space and time. In addition, the pressure dependent term $(\mu B)|_{x_{i\mp1/2}}$ of transmissibility uses some average viscosity and formation-volume-factor (FVF) of the fluid contained in Block i and Neighboring Block $i\mp1$ or some weight (up-stream weighting, average weighting) at any instant of time t. In other words, the term $(\mu B)|_{x_{i\mp1/2}}$ is not a function of space but it is a function of time as block pressures change with time. Similarly, for multi phase flow, the relative permeability of Phase $p = o, w, g$ between Block i and Neighboring Block $i\mp1$ at any instant of time t ($k_{rp}|_{x_{i\mp1/2}}$) uses upstream value or two-point upstream value of Block i and Neighboring Block $i\mp1$ that are already fixed in space. In other words, the term $k_{rp}|_{x_{i\mp1/2}}$ is not a function of space but it is a function of time as block saturations change with time. Hence, transmissibility $T_{x_{i\mp1/2}}$ between Block i and its Neighboring Block $i\mp1$ is a function of time only; it does not depend on space at any instant in time.

2. A close inspection of the flow terms on the LHS of the discretized flow equation expressed by Eq. 25 reveals that these terms are nothing but Darcy's law describing volumetric flow rates at standard conditions ($q_{sc_{i\mp1/2}}$) between Gridblock i and its neighboring Gridblocks i-1 and i+1 in the x-direction; i.e.,

$$T_{x_{i\mp1/2}}(p_{i\mp1}-p_i) = (\beta_c \frac{k_x A_x}{\mu B \Delta x})_{i\mp1/2}(p_{i\mp1}-p_i) = q_{sc_{i\mp1/2}}. \tag{37}$$

3. Interblock flow terms and production/injection rates that appear on the LHS of the discretized flow equations (Eqs. 29, 33, 36) are dated at Time Level n for explicit flow equation, Time Level $n+1$ for implicit flow equation, or Time Level $n+1/2$ for the Crank-Nicolson flow equation. In all cases, the RHS of the flow equations represent accumulation over a time step Δt. In other words, the accumulation term does not take into consideration the variation of interblock flow terms and production/injection rate (source/sink term) with time within a time step.

ENGINEERING APPROACH

In the engineering approach, the derivation of the algebraic flow equation is straight-forward. It is accomplished in three consecutive steps: 1) discretization of reservoir into gridblocks (or gridpoints) as shown earlier to remove the effect of space variable as mentioned in Observation 1 above, 2) derivation of the algebraic flow equation for Gridblock i (or Gridpoint i) using the three basic principles mentioned earlier taking into consideration the variation of interblock flow terms and source/sink term with time within a time step, and 3) approximation of the time integrals in the resulting flow equation to produce the nonlinear algebraic flow equations.

Derivation of the Algebraic Flow Equations

In the first step, the reservoir is discretized as mentioned earlier. Fig. 4 shows Gridblock i (or Gridpoint i), and its neighboring gridblocks in the x-direction (Gridblock i-1 and Gridblock i+1). At any instant in time, fluid enters Gridblock i, coming from Gridblock i-1, across its $x_{i-1/2}$ face at a mass rate of $w_x|_{x_{i-1/2}}$ and leaves to Gridblock i+1 across its $x_{i+1/2}$ face at a mass rate of $w_x|_{x_{i+1/2}}$. The fluid also enters Gridblock i through a well at a mass rate of q_m. The mass of fluid in Gridblock i per unit volume of rock is m_v.

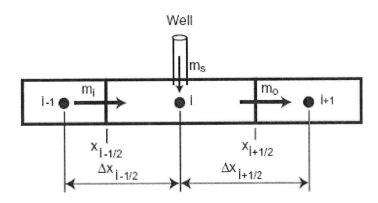

Figure 4. Gridblock i (or Gridpoint i) used for writing material balance in the engineering approach.

Therefore, the material balance equation written over a time step $\Delta t = t^{n+1} - t^n$ as expressed by Eq. 1 becomes

$$m_i\big|_{x_{i-1/2}} - m_o\big|_{x_{i+1/2}} + m_{s_i} = m_{a_i}. \qquad (38)$$

Terms like $w_x\big|_{x_{i-1/2}}$, $w_x\big|_{x_{i+1/2}}$, and q_{m_i} are functions of time only because space is not a variable for an already discretized reservoir (see Observation 1). Therefore,

$$m_i\big|_{x_{i-1/2}} = \int_{t^n}^{t^{n+1}} w_x\big|_{x_{i-1/2}} \, dt, \qquad (39)$$

$$m_o\big|_{x_{i+1/2}} = \int_{t^n}^{t^{n+1}} w_x\big|_{x_{i+1/2}} \, dt, \qquad (40)$$

and

$$m_{s_i} = \int_{t^n}^{t^{n+1}} q_{m_i} \, dt. \qquad (41)$$

Using Eqs. 39 through 41, Eq. 38 can be rewritten as

$$\int_{t^n}^{t^{n+1}} w_x\big|_{x_{i-1/2}} \, dt - \int_{t^n}^{t^{n+1}} w_x\big|_{x_{i+1/2}} \, dt + \int_{t^n}^{t^{n+1}} q_{m_i} \, dt = m_{a_i}. \qquad (42)$$

Substitution of Eq. 6 and Eq. 7 into Eq. 42 yields

$$\int_{t^n}^{t^{n+1}} (\dot{m}_x A_x)\big|_{x_{i-1/2}} \, dt - \int_{t^n}^{t^{n+1}} (\dot{m}_x A_x)\big|_{x_{i+1/2}} \, dt + \int_{t^n}^{t^{n+1}} q_{m_i} \, dt = V_{b_i} (m_v^{n+1} - m_v^n)_i. \qquad (43)$$

Substitution of Eq. 11 through 13 into Eq. 43 yields

$$\int_{t^n}^{t^{n+1}} (\alpha_c \rho u_x A_x)\big|_{x_{i-1/2}} \, dt - \int_{t^n}^{t^{n+1}} (\alpha_c \rho u_x A_x)\big|_{x_{i+1/2}} \, dt + \int_{t^n}^{t^{n+1}} (\alpha_c \rho q)_i \, dt = V_{b_i} [(\phi\rho)_i^{n+1} - (\phi\rho)_i^n]. \qquad (44)$$

Substitution of Eq. 2 into Eq. 44, dividing through by $\alpha_c \rho_{sc}$, and noting that $q/B = q_{sc}$ yields

$$\int_{t^n}^{t^{n+1}} (\frac{u_x A_x}{B})\bigg|_{x_{i-1/2}} dt - \int_{t^n}^{t^{n+1}} (\frac{u_x A_x}{B})\bigg|_{x_{i+1/2}} dt + \int_{t^n}^{t^{n+1}} q_{sc_i} dt = \frac{V_{b_i}}{\alpha_c}[(\frac{\phi}{B})_i^{n+1} - (\frac{\phi}{B})_i^n]. \quad (45)$$

Fluid volumetric velocity (flow rate per unit cross-sectional area) from Gridblock i-1 to Gridblock i is given by the algebraic analog of Eq. 3,

$$u_x\bigg|_{x_{i-1/2}} = \beta_c \frac{k_x|_{i-1/2}}{\mu} \frac{(p_{i-1} - p_i)}{\Delta x_{i-1/2}}. \quad (46)$$

Likewise, fluid flow rate per unit cross-sectional area from Gridblock i to Gridblock i+1 is

$$u_x\bigg|_{x_{i+1/2}} = \beta_c \frac{k_x|_{i+1/2}}{\mu} \frac{(p_i - p_{i+1})}{\Delta x_{i+1/2}}. \quad (47)$$

Substitution of Eqs. 46 and 47 into Eq. 45 and rearranging results in

$$\int_{t^n}^{t^{n+1}} [(\beta_c \frac{k_x A_x}{\mu B \Delta x})\bigg|_{x_{i-1/2}}(p_{i-1} - p_i)]dt - \int_{t^n}^{t^{n+1}} [(\beta_c \frac{k_x A_x}{\mu B \Delta x})\bigg|_{x_{i+1/2}}(p_i - p_{i+1})]dt + \int_{t^n}^{t^{n+1}} q_{sc_i} dt = \frac{V_{b_i}}{\alpha_c}[(\frac{\phi}{B})_i^{n+1} - (\frac{\phi}{B})_i^n] \quad (48)$$

or

$$\int_{t^n}^{t^{n+1}} [T_{x_{i-1/2}}(p_{i-1} - p_i)]dt + \int_{t^n}^{t^{n+1}} [T_{x_{i+1/2}}(p_{i+1} - p_i)]dt + \int_{t^n}^{t^{n+1}} q_{sc_i} dt = \frac{V_{b_i}}{\alpha_c}[(\frac{\phi}{B})_i^{n+1} - (\frac{\phi}{B})_i^n]. \quad (49)$$

The derivation of Eq. 49 is rigorous and involves no assumptions other than the validity of Darcy's law (Eqs. 46 and 47) to estimate fluid volumetric velocity between Gridblock i and its neighboring Gridblocks i-1 and i+1. Such validity is not questionable by petroleum engineers.

Again, the accumulation term in above equation can be expressed in terms of the pressure of Gridblock i; and Eq. 49 becomes

$$\int_{t^n}^{t^{n+1}} [T_{x_{i-1/2}}(p_{i-1} - p_i)]dt + \int_{t^n}^{t^{n+1}} [T_{x_{i+1/2}}(p_{i+1} - p_i)]dt + \int_{t^n}^{t^{n+1}} q_{sc_i} dt = \frac{V_{b_i}}{\alpha_c}(\frac{\phi}{B})_i'[p_i^{n+1} - p_i^n]. \quad (50)$$

Approximation of Time Integrals

A time integral such as those appearing in Eq. 49 or Eq. 50 is shown schematically in Fig. 5. If the argument of an integral is an explicit function of time, the integral can be evaluated analytically. This is not the case for the integrals appearing on the LHS of either Eq. 49 or Eq. 50.

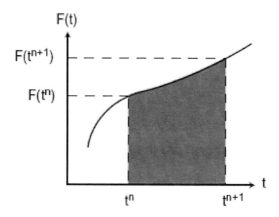

Figure 5. Representation of integral of a function as the area under the curve.

ODE Methods

If Eq. 50 is written for every Gridblock $i = 1, 2, 3...n_x$; then the solution can be obtained by one of the ODE methods (Euler's method, Modified Euler's method, the explicit Runge-Kutta method, the implicit Runge-Kutta method), and others reviewed and compared along with the explicit and implicit methods by Aziz and Settari (1979). They concluded that for the cases where the same PDE's are solved repeatedly for different problems, the greatest success is achieved by considering the peculiarities of the problem in developing simulation programs, which infers the exclusion of ODE methods. The explicit and implicit methods are presented in the next section.

Explicit and Implicit Methods

Performing the integrals on the LHS of Eq. 49 or Eq. 50 necessitates making certain assumptions. Such assumptions lead to deriving equations as those expressed by Eqs. 28, 32, and 35 (or Eqs. 29, 33, and 36).

Consider the integral $\int_{t^n}^{t^{n+1}} F(t) dt$ shown in Fig. 6. This integral can be evaluated as follows

$$\int_{t^n}^{t^{n+1}} F(t)dt \cong \int_{t^n}^{t^{n+1}} F(t^m)dt = \int_{t^n}^{t^{n+1}} F^m dt = F^m \int_{t^n}^{t^{n+1}} dt = F^m t \Big|_{t^n}^{t^{n+1}} = F^m(t^{n+1} - t^n) = F^m \Delta t. \quad (51)$$

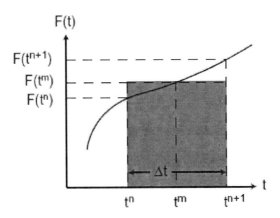

Figure 6. Representation of integral of function as $F(t^m) \times \Delta t$.

The argument F stands for $[T_{x_{i-1/2}}(p_{i-1} - p_i)]$, $[T_{x_{i+1/2}}(p_{i+1} - p_i)]$, or q_{sc_i} that appears on the LHS of Eq. 49 and F^m = approximation of F at time t^m = constant over the time interval Δt.

Forward-Central- Difference Equation

The forward-central-difference equation given by Eq. 28 can be obtained from Eq. 49 if the argument F of integrals is dated at time t^n; i.e., $F \cong F^m = F^n$ as shown in Fig. 7a. Therefore, Eq. 51 becomes $\int_{t^n}^{t^{n+1}} F(t) dt \cong F^n \Delta t$, and Eq. 49 reduces to

$$[T^n_{x_{i-1/2}}(p^n_{i-1} - p^n_i)]\Delta t + [T^n_{x_{i+1/2}}(p^n_{i+1} - p^n_i)]\Delta t + q^n_{sc_i}\Delta t \cong \frac{V_{b_i}}{\alpha_c}[(\frac{\phi}{B})^{n+1}_i - (\frac{\phi}{B})^n_i]. \quad (52)$$

Dividing above equation by Δt gives Eq. 28,

$$T^n_{x_{i-1/2}}(p^n_{i-1} - p^n_i) + T^n_{x_{i+1/2}}(p^n_{i+1} - p^n_i) + q^n_{sc_i} \cong \frac{V_{b_i}}{\alpha_c \Delta t}[(\frac{\phi}{B})^{n+1}_i - (\frac{\phi}{B})^n_i]. \quad (28)$$

If one starts with Eq. 50 instead of Eq. 49, he ends up with Eq. 29,

$$T^n_{x_{i-1/2}}(p^n_{i-1} - p^n_i) + T^n_{x_{i+1/2}}(p^n_{i+1} - p^n_i) + q^n_{sc_i} \cong \frac{V_{b_i}}{\alpha_c \Delta t}(\frac{\phi}{B})'_i[p^{n+1}_i - p^n_i]. \quad (29)$$

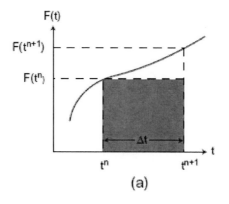

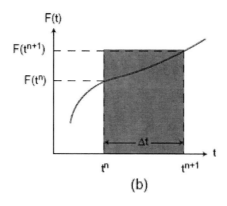

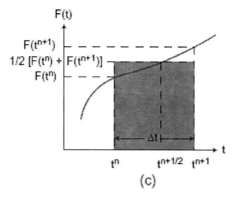

Figure 7. Different methods of approximation of the integral of a function.

Backward-Central-Difference Equation

The backward-central-difference equation given by Eq. 32 can be obtained from Eq. 49 if the argument F of integrals is dated at time t^{n+1}; i.e., $F \cong F^m = F^{n+1}$ as shown in Fig. 7b.

Therefore, Eq. 51 becomes $\int_{t^n}^{t^{n+1}} F(t)\,dt \cong F^{n+1}\Delta t$, and Eq. 49 reduces to

$$[T^{n+1}_{x_{i-1/2}}(p^{n+1}_{i-1}-p^{n+1}_{i})]\Delta t+[T^{n+1}_{x_{i+1/2}}(p^{n+1}_{i+1}-p^{n+1}_{i})]\Delta t+q^{n+1}_{sc_i}\Delta t \cong \frac{V_{b_i}}{\alpha_c}[(\frac{\phi}{B})^{n+1}_i-(\frac{\phi}{B})^{n}_i] \tag{53}$$

Dividing above equation by Δt gives Eq. 32,

$$T^{n+1}_{x_{i-1/2}}(p^{n+1}_{i-1}-p^{n+1}_{i})+T^{n+1}_{x_{i+1/2}}(p^{n+1}_{i+1}-p^{n+1}_{i})+q^{n+1}_{sc_i} \cong \frac{V_{b_i}}{\alpha_c \Delta t}[(\frac{\phi}{B})^{n+1}_i-(\frac{\phi}{B})^{n}_i]. \tag{32}$$

If one starts with Eq. 50 instead of Eq. 49, he ends up with Eq. 33,

$$T^{n+1}_{x_{i-1/2}}(p^{n+1}_{i-1}-p^{n+1}_{i})+T^{n+1}_{x_{i+1/2}}(p^{n+1}_{i+1}-p^{n+1}_{i})+q^{n+1}_{sc_i} \cong \frac{V_{b_i}}{\alpha_c \Delta t}(\frac{\phi}{B})'_i[p^{n+1}_i-p^{n}_i]. \tag{33}$$

Second-Order-Central Difference (Crank-Nicholson) Equation

The second order in time Crank-Nicholson approximation given by Eq. 35 can be obtained from Eq. 49 if the argument F of integrals is dated at time $t^{n+1/2}$. This choice of time level was made in order to make the RHS of Eq. 26 to appear as second order approximation in time in the mathematical approach. In this case, the argument F in the integrals can be approximated as $F \cong F^m = F^{n+1/2} = (F^n + F^{n+1})/2$ as shown in Fig. 7c. Therefore, Eq. 51 becomes $\int_{t^n}^{t^{n+1}} F(t)dt \cong \frac{1}{2}(F^n + F^{n+1})\Delta t$, and Eq. 49 reduces to

$$(1/2)[T^{n}_{x_{i-1/2}}(p^{n}_{i-1}-p^{n}_{i})+T^{n+1}_{x_{i-1/2}}(p^{n+1}_{i-1}-p^{n+1}_{i})]\Delta t+(1/2)[T^{n}_{x_{i+1/2}}(p^{n}_{i+1}-p^{n}_{i})+T^{n+1}_{x_{i+1/2}}(p^{n+1}_{i+1}-p^{n+1}_{i})]\Delta t$$

$$+(1/2)[q^{n+1}_{sc_i}+q^{n}_{sc_i}]\Delta t \cong \frac{V_{b_i}}{\alpha_c}[(\frac{\phi}{B})^{n+1}_i-(\frac{\phi}{B})^{n}_i]. \tag{54}$$

Dividing above equation by Δt and rearranging terms gives Eq. 35,

$$(1/2)[T^{n}_{x_{i-1/2}}(p^{n}_{i-1}-p^{n}_{i})+T^{n}_{x_{i+1/2}}(p^{n}_{i+1}-p^{n}_{i})]+(1/2)[T^{n+1}_{x_{i-1/2}}(p^{n+1}_{i-1}-p^{n+1}_{i})+T^{n+1}_{x_{i+1/2}}(p^{n+1}_{i+1}-p^{n+1}_{i})]$$

$$+(1/2)[q^{n}_{sc_i}+q^{n+1}_{sc_i}] \cong \frac{V_{b_i}}{\alpha_c \Delta t}[(\frac{\phi}{B})^{n+1}_i-(\frac{\phi}{B})^{n}_i]. \tag{35}$$

If one starts with Eq. 50 instead of Eq. 49, he ends up with Eq. 36,

$$(1/2)[T^{n}_{x_{i-1/2}}(p^{n}_{i-1}-p^{n}_{i})+T^{n}_{x_{i+1/2}}(p^{n}_{i+1}-p^{n}_{i})]+(1/2)[T^{n+1}_{x_{i-1/2}}(p^{n+1}_{i-1}-p^{n+1}_{i})+T^{n+1}_{x_{i+1/2}}(p^{n+1}_{i+1}-p^{n+1}_{i})]$$

$$+(1/2)[q_{sc_i}^n + q_{sc_i}^{n+1}] \cong \frac{V_{b_i}}{\alpha_c \Delta t}(\frac{\phi}{B})'_i[p_i^{n+1} - p_i^n]. \tag{36}$$

Therefore, one can conclude that the same nonlinear algebraic equations can be derived by the mathematical and engineering approaches.

Treatment of Initial and Boundary Conditions

Initial condition receives the same treatment by both mathematical and engineering approaches. Therefore, this section focuses on the treatment of boundary conditions by both approaches and highlights differences. An external (or internal) reservoir boundary can be subject to one of three conditions: no-flow boundary, constant flow boundary, or constant pressure boundary. In fact, the first two boundary conditions reduce to specified pressure gradient condition (Neumann boundary condition) and the third boundary condition is the Dirichlet boundary condition. The treatment of boundary conditions by the engineering approach is presented first, followed by comparison with the treatment by the mathematical approach.

The engineering approach replaces any boundary condition by a no-flow boundary condition plus a fictitious well. The flow rate of the fictitious well ($q_{sc_{b,bB}}$) depends on the specified boundary condition and is calculated using Darcy's law between the boundary (b) and the point representing the Boundary Block bB.

Specified Boundary Pressure

$$q_{sc_{b,bB}}^m = \frac{1}{\Delta t}\int_{t^n}^{t^{n+1}}[T_{b,bB}(p_b - p_{bB})]dt = \frac{1}{\Delta t}\int_{t^n}^{t^{n+1}}[T_{b,bB}^m(p_b - p_{bB}^m)]dt$$

$$= [T_{b,bB}^m(p_b - p_{bB}^m)]\int_{t^n}^{t^{n+1}}dt/\Delta t \tag{55}$$

or

$$q_{sc_{b,bB}}^m = T_{b,bB}^m(p_b - p_{bB}^m), \tag{56}$$

where $q_{sc_{b,bB}}^m$ = flow rate across reservoir boundary into Boundary Gridblock bB,

$$T_{b,bB}^m = (\beta_c \frac{k_x A_x}{\mu B \Delta x})_{b,bB}^m \tag{57}$$

represents fluid transmissibility between the reservoir boundary and the point representing Boundary Gridblock bB, and Superscript m reflects time level approximation mentioned earlier. In the following presentation, we demonstrate the treatment of boundary conditions at $x = 0$.

For point-distributed grid (see Fig. 8a), $p_1 = p_b$, $p_2 = p_{bP}$, and $T_{1+1/2} = T_{b,bP}$. Substitution of these relations into Eq. 56 gives,

$$q_{sc_{b,2}} = T_{1+1/2}(p_1 - p_2), \qquad (58)$$

which is the inter-block flow rate ($q_{sc_{1+1/2}}$) between Gridpoints 1 and 2 as given by the mathematical approach for the chosen time level approximation.

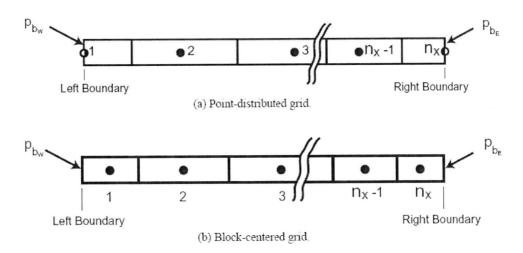

Figure 8. Dirichlet boundary condition for: (a) point-distributed grid, (b) block-centered grid.

For block-centered grid, application of Eq. 56 gives

$$q_{sc_{b,1}} = T_{b,1}(p_b - p_1) = [\beta_c \frac{k_x A_x}{(\mu B \Delta x_1 / 2)}]_1 (p_b - p_1). \qquad (59)$$

In comparison, the mathematical approach (see Fig. 8b) assumes $p_1 \cong p_b$ (first order approximation). The resulting flow rate crossing reservoir boundary is given by the inter-block flow rate between Gridblocks 1 and 2; i.e.,

$$q_{sc_{1+1/2}} = T_{1+1/2}(p_1 - p_2) = T_{1+1/2}(p_b - p_2). \qquad (60)$$

If a second order approximation is used (Aziz and Settari, 1979), then

$$p_b \cong \frac{\Delta x_{1/2} + \Delta x_{1+1/2}}{\Delta x_{1+1/2}} p_1 - \frac{\Delta x_{1/2}}{\Delta x_{1+1/2}} p_2. \tag{61}$$

The fluid flow equation for Gridblock 1 is written as

$$T_{x_{1/2}}(p_0 - p_1) + T_{x_{1+1/2}}(p_2 - p_1) + q_{sc_1} = \frac{V_{b_1}}{\alpha_c \Delta t}[(\frac{\phi}{B})_1^{n+1} - (\frac{\phi}{B})_1^{n}], \tag{62}$$

The first term on the LHS of Eq. 62 can be rewritten as

$$T_{x_{1/2}}(p_0 - p_1) = T_{x_{1/2}}[(p_0 - p_b + p_b - p_1)] = T_{x_{1/2}}(p_0 - p_b) + T_{x_{1/2}}(p_b - p_1). \tag{63}$$

In order to keep the pressure on the left boundary of Gridblock 1 constant, the fluid leaving the boundary to one side (e.g., Gridpoint 1) has to be equal to the fluid entering the boundary from the other side (e.g., Gridpoint 0); i.e.,

$$T_{x_{1/2}}(p_0 - p_b) = T_{x_{1/2}}(p_b - p_1). \tag{64}$$

In addition, note that

$$T_{x_{1/2}} = [\beta_c \frac{k_x A_x}{\mu B \Delta x}]_{1/2} = [\beta_c \frac{k_x A_x}{\mu B \Delta x_1}]_1 = \frac{1}{2}[\beta_c \frac{k_x A_x}{(\mu B \Delta x_1/2)}]_1 = \frac{1}{2}T_{b,1}. \tag{65}$$

Substitution of Eqs. 64 and 65 into Eq. 63 results in

$$T_{x_{1/2}}(p_0 - p_1) = 2T_{x_{1/2}}(p_b - p_1) = T_{b,1}(p_b - p_1) = q_{sc_{b,1}}, \tag{66}$$

where $q_{sc_{b,1}}$ is nothing but the fictitious flow rate for the Boundary Gridblock 1 expressed by Eq. 59 for the engineering approach.

Specified Flow Rate Across Boundary

$$q_{sc_{b,bB}} = \frac{1}{\Delta t}\int_{t^n}^{t^{n+1}} q_{spsc} dt = \frac{1}{\Delta t} q_{spsc} \int_{t^n}^{t^{n+1}} dt = \frac{q_{spsc}\Delta t}{\Delta t} = q_{spsc}, \tag{67}$$

where q_{spsc} = specified flow rate (converted to standard conditions) across reservoir boundary, $q_{spsc} \neq 0$ reflects constant flow rate boundary condition, and $q_{spsc} = 0$ reflects no-flow boundary condition.

If pressure gradient at reservoir boundary is specified (see Fig. 9), then

$$q_{sc_{b,bB}} = (\beta_c \frac{k_x A_x}{\mu B})_{bB} \frac{\partial p}{\partial x}\bigg|_b , \qquad (68)$$

where $q_{sc_{b,bB}}$ is positive for injection and negative for production. Applying Eq. 68 for a boundary gridblock (or gridpoint) gives:

$$q_{sc_{b,1}} = -(\beta_c \frac{k_x A_x}{\mu B})_1 \frac{\partial p}{\partial x}\bigg|_b \qquad (69)$$

for Gridblock 1 (or Gridpoint 1) because injection creates negative pressure gradient, while production creates positive pressure gradient at this boundary; and

$$q_{sc_{b,n_x}} = +(\beta_c \frac{k_x A_x}{\mu B})_{n_x} \frac{\partial p}{\partial x}\bigg|_b \qquad (70)$$

for Gridblock n_x (or Gridpoint n_x) because injection creates positive pressure gradient, while production creates negative pressure gradient at this boundary.

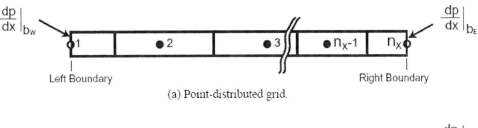

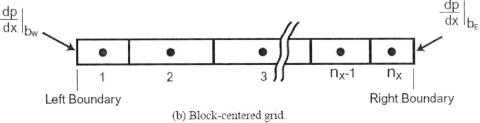

Figure 9. Neumann boundary condition for: (a) point-distributed grid, (b) block-centered grid.

For the mathematical approach, we will demonstrate the application of boundary pressure gradient specification to Gridblock 1 and Gridpoint 1. A second order approximation for the pressure gradient is possible using the 'reflection technique' and by introducing an auxiliary point (p_0) outside the reservoir on the other side of the boundary as shown in Fig. 10. Aziz and Settari (1979) have reported the discretization of this boundary condition for both block-centered and point-distributed grids for regular grids. The discretization of this boundary condition is presented here for irregular grids.

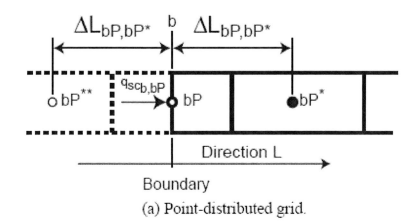

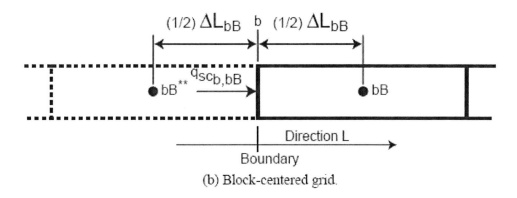

Figure 10. Reflection technique for: (a) point-distributed grid, (b) block-centered grid.

For point-distributed grid (Fig. 10a),

$$\left.\frac{\partial p}{\partial x}\right|_b \cong \frac{p_{bP^*} - p_{bP^{**}}}{2\Delta L_{bP,bP^*}} = \frac{p_2 - p_0}{2\Delta x_{1+1/2}}. \tag{71}$$

The difference equation for the whole boundary block represented by Gridpoint 1 ($bP = 1$) is

$$T_{x_{1/2}}(p_0 - p_1) + T_{x_{1+1/2}}(p_2 - p_1) + 2q_{sc_1} = \frac{2V_{b_1}}{\alpha_c \Delta t}[(\frac{\phi}{B})_1^{n+1} - (\frac{\phi}{B})_1^n], \quad (72)$$

because $V_b = 2V_{b_1}$ and $q_{sc} = 2q_{sc_1}$. Using Eq. 71 to eliminate p_0 from Eq. 72, dividing the resulting equation by 2, and observing that $\Delta x_{1/2} = \Delta x_{1+1/2}$ and $T_{x_{1/2}} = T_{x_{1+1/2}}$ because of the reflection technique, one obtains

$$-T_{x_{1/2}} \Delta x_{1/2} \frac{\partial p}{\partial x}\bigg|_b + T_{x_{1+1/2}}(p_2 - p_1) + q_{sc_1} = \frac{V_{b_1}}{\alpha_c \Delta t}[(\frac{\phi}{B})_1^{n+1} - (\frac{\phi}{B})_1^n]. \quad (73)$$

Note that the first term on the LHS of above equation is nothing but $q_{sc_{b,1}}$ given by Eq. 69. Therefore, Eq. 73 reduces to

$$q_{sc_{b,1}} + T_{x_{1+1/2}}(p_2 - p_1) + q_{sc_1} = \frac{V_{b_1}}{\alpha_c \Delta t}[(\frac{\phi}{B})_1^{n+1} - (\frac{\phi}{B})_1^n]. \quad (74)$$

This is exactly the equation for Gridpoint 1 that will be produced by the engineering approach.

For block-centered grid (Fig. 10b),

$$\frac{\partial p}{\partial x}\bigg|_b \cong \frac{p_{bB} - p_{bB^{**}}}{\Delta L_{bB}} = \frac{p_1 - p_0}{\Delta x_1}. \quad (75)$$

The difference equation for Gridblock 1 ($bB = 1$) is

$$T_{x_{1/2}}(p_0 - p_1) + T_{x_{1+1/2}}(p_2 - p_1) + q_{sc_1} = \frac{V_{b_1}}{\alpha_c \Delta t}[(\frac{\phi}{B})_1^{n+1} - (\frac{\phi}{B})_1^n]. \quad (76)$$

Using Eq. 75 to eliminate p_0 from Eq. 76, one obtains Eq. 73. Note that the first term on the LHS of Eq. 73 is nothing but $q_{sc_{b,1}}$ given by Eq. 69 because $\Delta x_{1/2} = \Delta x_1$ and $T_{x_{1/2}} = T_{x_1}$ because of the reflection technique. Therefore, Eq. 74 becomes the final finite-difference equation for Gridblock 1. This is exactly the equation for Gridblock 1 that will be produced by the engineering approach.

From the above analysis of treatment of boundary conditions, one can conclude that both the engineering and mathematical approaches give identical equations for specified flow rate (or pressure gradient) boundary condition and specified pressure boundary condition for both the point-distributed and block-centered grids. This treatment of boundary conditions is second order correct. The engineering approach gives more accurate treatment than the mathematical approach if first order approximation is used in the treatment of specified pressure boundary condition in block-centered grid ($p_1 \cong p_b$) (Abou-Kassem and Osman, 2006). Using the engineering approach, Abou-Kassem, Farouq Ali, and Islam (2006) derived the flow equations and presented the treatment of boundary conditions for the cases of single-well simulation in radial-cylindrical coordinates and multi-dimensional, multi-phase flow in black-oil models.

POSSIBLE REASONS FOR OVERLOOKING THE ENGINEERING APPROACH

The engineering approach has been overlooked all these years. Traditionally, reservoir simulators were developed by first using a control volume (or elementary volume), that was envisaged by mathematicians [see Fig. 3 for 1D reservoir (Aziz and Settari, 1979) or Fig. 11 for 3D reservoir (Bear, 1988)] to develop fluid flow equations. The resulting flow equations are in the form of PDE's. Once the PDE's are derived, early pioneers of simulation were looking up to mathematics to provide solution methods. These methods of solution, as mentioned in the introduction, start with the description of the reservoir as a collection of gridblocks, represented by points that fall within them (see Fig. 2), followed by the replacement of the PDE's by algebraic equations, and finally solving the resulting algebraic equations. The simulator developer was all the time occupied by finding the solution and, perhaps, forgot that he was solving an engineering problem with blocks similar to those shown in Fig. 2. The engineering approach can be realized should one try to relate the discretized flow equations for Gridblock i to all gridblocks (or gridpoints) shown in Fig. 2. A close inspection of the flow terms on the LHS of a discretized flow equation reveals that these terms are nothing but Darcy's law describing volumetric flow rates at standard conditions ($q_{sc_{i\mp1/2}}$) between Gridblock i and its neighboring Gridblocks i-1 and i+1 in the x-direction (see Observation 2); i.e.,

$$T_{x_{i\mp1/2}}(p_{i\mp1} - p_i) = (\beta_c \frac{k_x A_x}{\mu B \Delta x})_{i\mp1/2}(p_{i\mp1} - p_i) = q_{sc_{i\mp1/2}}. \tag{37}$$

Farouq Ali (1986) was the first to observe this relationship more than 30 years ago. Making use of this observation, he developed the forward-central-difference equation (Eq. 28) and the backward-central-difference equation (Eq. 32) without going through the rigor of the mathematical approach in teaching reservoir simulation to undergraduates. Ertekin, Abou-Kassem, and King (2001) were also the first to use a control volume with the Point x at its center that is closer to engineer's thinking (Fig. 12 for 1D reservoir and Fig. 13 for 3D reservoir) in the mathematical approach. The observation by Farouq Ali in the early seventies and the introduction of the new control volume by Ertekin et al. were the two milestones that contributed significantly to the development of the engineering approach.

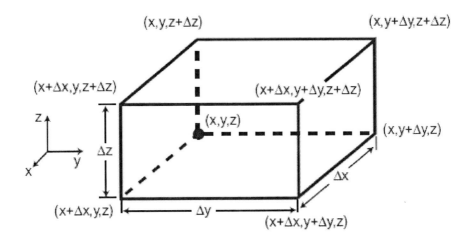

Figure 11. Control volume in 3D traditionally used for writing material balance. [Redrawn from Bear (1988)]

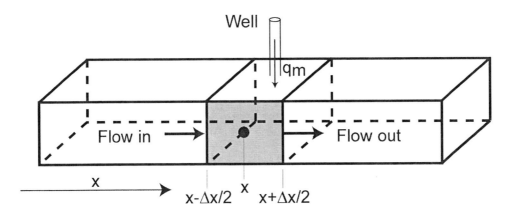

Figure 12. Control volume in 1D used for writing material balance. [Redrawn from Ertekin et al. (2001)]

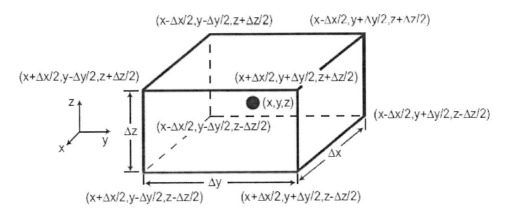

Figure 13. Control volume in 3D used for writing material balance. [Redrawn from Ertekin et al. (2001)]

Overlooking the engineering approach has kept reservoir simulation closely tied with PDE's. From mathematician's point of view, it is a blessing because researchers in reservoir simulation have devised advanced methods for solving highly nonlinear PDE's, and this enriched the literature in mathematics in this important area. Contributions of reservoir simulation to solving PDE's include:

1) treatment of nonlinear terms in space and time (Settari and Aziz, 1975; Coats, Ramesh, and Winestock, 1977; Saad, 1989; Gupta, 1990),
2) devising methods of solving systems of nonlinear PDE's: IMPES (Breitenbach, Thurnau, and van Poolen, 1969), SEQ (Spillette, Hillestad, and Stone, 1973; Coats, 1978), Fully Implicit SS (Sheffield, 1969), and Adaptive Implicit (Thomas and Thurnau, 1983) Methods, and
3) devising advanced iterative methods for solving systems of linear algebraic equations: Block Iterative (Behie and Vinsome, 1982), Nested Factorization (Appleyard and Cheshire, 1983), and Orthomin (Vinsome, 1976).

IMPORTANCE OF THE ENGINEERING APPROACH AND THE MATHEMATICAL APPROACH

The importance of the engineering approach lies in being close to the engineer's thinking and in its justification of the central-difference approximation of the second order space derivative used in the mathematical approach. It also provides the physical interpretation of the approximations involved in the forward-, backward-, and central-difference of the first order time derivative used in the mathematical approach. This is so because the engineering approach is independent of the mathematical approach. In addition, the algebraic equations are easily obtained without going through the rigor of the mathematical approach. In reality, the development of a reservoir simulator can do away with the mathematical approach because the objective is to obtain the appropriate nonlinear algebraic equations for the process being simulated. The majority of, if not all, available commercial reservoir simulators were developed without even looking at analysis of truncation errors, consistency, convergence, or stability. The importance of the mathematical approach; however, lies in its capacity to provide analysis of such items. Only then, the two approaches complement each other and both become equally important in reservoir simulation.

CONCLUSION

The following conclusions can be drawn:

1. The discretized flow equations (nonlinear algebraic equations) in reservoir simulation of any process in Cartesian and radial-cylindrical coordinate systems can be obtained in a rigorous way by the engineering approach without going through the rigor of obtaining the PDE's describing the process and the space and time discretization (mathematical approach).
2. The engineering approach is closer to engineer's thinking than mathematical approach. While the mathematical approach derives the nonlinear algebraic equations

by first deriving the PDE's, followed by discretizing the reservoir, and finally discretizing the PDE's; the engineering approach first discretizes the reservoir, then derives the algebraic equations with time integrals, and finally approximates the time integrals to obtain the nonlinear algebraic equations.
3. Both the engineering and mathematical approaches treat boundary conditions with the same accuracy if second order approximation is used. If discretization of specified boundary pressure condition in block-centered grid is first order correct, then the engineering approach gives a representation that is more accurate.
4. The engineering approach is closer to the physical meaning of the equations and provides physical justification for using central-difference approximation of flow terms and give interpretation of the forward-, backward, and central-difference approximations of the accumulation term. Analysis of local truncation errors, consistency, convergence, and stability; however, can be studied by the mathematical approach only. Therefore, the mathematical and engineering approaches complement each other.
5. The ODE methods such as the implicit Runge-Kutta method should be reconsidered because it has the capacity to perform the integrals over a time step in Eqs. 49 and 50.

NOMENCLATURE

A_x =cross-sectional area normal to x-direction, ft² [m²]

$A_x|_x$ =cross-sectional area normal to x-direction at Point x, ft² [m²]

$A_x|_{x+\Delta x}$ =cross-sectional area normal to x-direction at Point $x + \Delta x$, ft² [m²]

$A_x|_{x_{i\mp 1/2}}$ =cross-sectional area normal to x-direction at Block Boundary $x_{i\mp 1/2}$, ft² [m²]

B =fluid formation volume factor, RB/STB or RB/scf [m³/std m³]

k_x =permeability in the x-direction, md [μm²]

$k_x|_{i\mp 1/2}$ =permeability in the x-direction between Gridblock i and Gridblock $i \mp 1$, md [μm²]

L_x =reservoir length in the x-direction, ft [m]

m_a =mass accumulation, lbm [kg]

m_{a_i} =mass accumulation in Gridblock i, lbm [kg]

m_i =mass of fluid entering reservoir from other parts of reservoir, lbm [kg]

$m_i|_x$ =mass of fluid entering control volume boundary at x, lbm [kg]

$m_i|_{x_{i-1/2}}$ =mass of fluid entering Gridblock i Boundary $x_{i-1/2}$, lbm [kg]

m_o =mass of fluid leaving reservoir to other parts of reservoir, lbm [kg]

$m_o|_{x+\Delta x}$ =mass of fluid leaving control volume boundary at $x + \Delta x$, lbm [kg]

$m_o|_{x_{i+1/2}}$ = mass of fluid leaving Gridblock i Boundary $x_{i+1/2}$, lbm [kg]

m_s = mass of fluid entering (or leaving) reservoir through a well, lbm [kg]

m_{s_i} = mass of fluid entering (or leaving) Gridblock i through a well, lbm [kg]

m_v = mass of fluid per unit volume of reservoir rock, lbm/ft^3 [kg/m^3]

m_v^n = mass of fluid per unit volume of reservoir rock at Time t^n, lbm/ft^3 [kg/m^3]

m_v^{n+1} = mass of fluid per unit volume of reservoir rock at Time t^{n+1}, lbm/ft^3 [kg/m^3]

$m_v|_t$ = mass of fluid per unit volume of reservoir rock at Time t, lbm/ft^3 [kg/m^3]

$m_v|_{t+\Delta t}$ = mass of fluid per unit volume of reservoir rock at Time $t + \Delta t$, lbm/ft^3 [kg/m^3]

$\dot{m}_x$ = mass flux, lbm/D-ft^2 [kg/(d.m^2)]

$\dot{m}_x|_x$ = mass flux across control volume boundary at x, lbm/D-ft^2 [kg/(d.m^2)]

$\dot{m}_x|_{x+\Delta x}$ = mass flux across control volume boundary at $x + \Delta x$, lbm/D-ft^2 [kg/(d.m^2)]

$\dot{m}_x|_{x_{i\mp 1/2}}$ = mass flux across Gridblock i Boundary $x_{i\mp 1/2}$, lbm/D-ft^2 [kg/(d.m^2)]

n_x = number of reservoir gridblocks (or gridpoints) in the x-direction

w_x = mass rate, lbm/D [kg/d]

$w_x|_x$ = mass rate entering control volume boundary at x, lbm/D [kg/d]

$w_x|_{x+\Delta x}$ = mass rate leaving control volume boundary at $x + \Delta x$, lbm/D [kg/d]

$w_x|_{x_{i\mp 1/2}}$ = mass rate entering (or leaving) Gridblock i Boundary $x_{i\mp 1/2}$, lbm/D [kg/d]

p = pressure, psia [kPa]

p_i = pressure of Gridblock i, psia [kPa]

p_i^n = pressure of Gridblock i at Time t^n, psia [kPa]

p_i^{n+1} = pressure of Gridblock i at Time t^{n+1}, psia [kPa]

p_{i-1} = pressure of Gridblock i-1, psia [kPa]

p_{i-1}^n = pressure of Gridblock i-1 at Time t^n, psia [kPa]

p_{i-1}^{n+1} = pressure of Gridblock i-1 at Time t^{n+1}, psia [kPa]

p_{i+1} = pressure of Gridblock i+1, psia [kPa]

p_{i+1}^n = pressure of Gridblock i+1 at Time t^n, psia [kPa]

p_{i+1}^{n+1} = pressure of Gridblock i+1 at Time t^{n+1}, psia [kPa]

q = well volumetric rate at reservoir conditions, RB/D [m³/d]

q_m = mass rate entering control volume through a well, lbm/D [kg/d]

q_{sc} = well volumetric rate at standard conditions, STB/D or scf/D [std m³/d]

q_{sc_i} = well volumetric rate at standard conditions in Gridblock i, STB/D or scf/D [std m³/d]

$q_{sc_{i\mp1/2}}$ = inter-block volumetric flow rates at standard conditions between Gridblock i and Gridblock $i\mp1$, STB/D or scf/D [std m³/d]

$q_{sc_{b,bB}}$ = specified volumetric flow rate at standard conditions across reservoir boundary of Gridblock bB (or Gridpoint bB), STB/D or scf/D [std m³/d]

$q_{sc_{b,1}}$ = specified volumetric flow rate at standard conditions across reservoir boundary of Gridblock 1 (or Gridpoint 1), STB/D or scf/D [std m³/d]

t = time, day

t^n = old time level, day

t^{n+1} = new or current time level, day

$T_{x_{i\mp1/2}}$ = transmissibility, STB/D-psi or scf/D-psi [std m³/(d.kPa)]

u_x = volumetric velocity in the x-direction, RB/D-ft² [m³/(d.kPa)]

x = distance in the x-direction in the Cartesian coordinate system, ft [m]

x_i = coordinate of Gridblock i the x-direction, ft [m]

$x_{i\mp1}$ = coordinate of Gridblock $i\mp1$ the x-direction, ft [m]

$x_{i\mp1/2}$ = coordinate of Gridblock i Boundary $x_{i\mp1/2}$ in the x-direction, ft [m]

x_{n_x} = x-direction coordinate of Gridblock n_x, ft [m]

V_b = bulk volume of control volume, ft³ [m³]

V_{b_i} = bulk volume of Gridblock i, ft³ [m³]

$(\frac{\partial p}{\partial x})_{i\mp1/2}$ = pressure gradient in the x-direction evaluated at Gridblock i Boundary $x_{i\mp1/2}$, psi/ft [kPa/m]

α_c = volume conversion factor = 5.614583 for Customary Units or 1 for SPE Preferred SI Units

β_c = transmissibility conversion factor = 0.001127 for Customary Units or 0.0864 for SPE Preferred SI Units

Δx = size of control volume in the x-direction, ft [m]

Δx_i = size of Gridblock i in the x-direction, ft [m]

$\Delta x_{i\mp1}$ = size of Gridblock $i\mp1$ in the x-direction, ft [m]

$\Delta x_{i\mp1/2}$ = distance between Gridblock i and Gridblock $i\mp1$ in the x-direction, ft [m]

ϕ = porosity, fraction

μ = fluid viscosity, cp [Pa.s]

ρ = fluid density at reservoir conditions, lbm/ft^3 [kg/m^3]

ρ_{sc} = fluid density at standard conditions, lbm/ft^3 [kg/m^3]

ACKNOWLEDGEMENTS

The author would like to thank Mr. Othman N. Matahen, of the UAEU Center of Teaching and Learning Technology, for producing the drawings in this paper.

REFERENCES

Abou-Kassem, J.H., Farouq Ali, S.M., and Islam, M.R.: *Petroleum Reservoir Simulaton: A Basic Approach*, Gulf Publishing Company, Houston, TX. (2006).

Abou-Kassem, J.H. and Osman, M.E.: "An engineering approach to the treatment of constant pressure boundary condition in block-centered grid in reservoir simulation," *J. Pet. Sci. Tech.* (in press).

Appleyard, J.R. and Cheshire, I.M.: "Nested factorization," SPE # 12264, 1983 SPE Reservoir Simulation Symposium, San Francisco, 15-18 November.

Aziz, K. and Settari, A.: *Petroleum Reservoir Simulation*, Applied Science Publishers Ltd., London. (1979).

Bear, J.: *Dynamics of Fluids in Porous Media*, Dover Publications Inc., New York City, p. 196. (1988).

Behie, A. and Vinsome, P.K.W.: "Block iterative methods for fully implicit reservoir simulation," *SPEJ*, 22 (5): pp. 658-668. (1982).

Breitenbach, E.A., Thurnau, D.H., and van Poolen, H.K.: "The fluid flow simulation equations," *SPEJ*, 9 (2): pp. 155-169. (1969).

Coats, K.H., Ramesh, A.B., and Winestock, A.G.: "Numerical modeling of thermal reservoir behavior," Canada-Venezuela Oil Sands Symposium, Edmonton, Alberta, May 27-June 4, 1977.

Ertekin, T., Abou-Kassem, J.H., and King, G.R.: *Basic Applied Reservoir Simulation*, SPE Textbook Series Vol. 7, Richardson, Texas. (2001).

Farouq Ali, S.M.: *Elements of Reservoir Modeling and Selected Papers*, course notes, Department of Mineral Engineering, U. of Alberta, Alberta, Canada. (1986).

Gupta, A.D.: "Accurate resolution of physical dispersion in multidimensional numerical modeling of miscible and chemical displacement," *SPERE*, 5 (4): pp. 581-588. (1990).

Odeh, A.S.: "An Overview of Mathematical Modeling of the Behavior of Hydrocarbon Reservoirs," *SIAM Rev.*, 24 (3), 263. (1982).

Saad, N.: "Field scale simulation of chemical flooding," PhD dissertation, U. of Texas, Austin. (1989).

Sheffield, M.: "Three phase flow including gravitational, viscous, and capillary forces," *SPEJ*, 9 (3): pp. 255-269. (1969).

Settari, A. and Aziz, K.: "Treatment of nonlinear terms in the numerical solution of partial differential equations for multiphase flow in porous media," *Int. J. Multiphase Flow*, 1: 817-844. (1975).

Spillette, A.G., Hillestad, J.G., and Stone, H.L.: "A high-stability sequential solution approach to reservoir simulation," SPE # 4542, 48th Annual Fall Meeting, Las Vegas, Nevada, 30 Sept.- 3 Oct. (1973).

Thomas, G.W. and Thurnau, D.H.: "Reservoir simulation using an adaptive implicit method," *SPEJ*, 23 (5): pp. 759-768. (1983).

Vinsome, P.K.W.: "Orthomin, an iterative method for solving sparse banded sets of simultaneous linear equations," SPE # 5729, 1976 SPE Symposium on Numerical Simulation of Reservoir Performance, Los Angeles, 19-20 February.

In: Nature Science and Sustainable Technology
Editor: M. R. Islam, pp. 73-82

ISBN: 978-1-60456-009-1
© 2008 Nova Science Publishers, Inc.

Chapter 4

SURFACE CHEMISTRY OF ATLANTIC COD SCALE

A. Basu[a,1], R. L. White[b], M. D. Lumsden[b], P. Bishop[a], S. Butt[a], S. Mustafiz[*a] and M. R. Islam[a]

[a]Faculty of Engineering
[b]Department of Chemistry
Dalhousie University, Halifax NS Canada B3J 2X4
[1]Current affiliation: Department of Agricultural and
Biological Engineering, University of Idaho,
Moscow, ID, 83843, USA

ABSTRACT

Binding of heavy metal ions to Atlantic Cod (Gadus Morhua) scale may be utilized for purification of metal bearing effluents. To correctly assess the adsorption behavior, Ultimate analysis, SEM and NMR spectroscopic techniques were utilized to study the molecular structure of Atlantic Cod scale. Calcium and phosphorus components and adsorbed metallic elements were indicated by SEM techniques. The protein component indicated by the carbon to nitrogen ratio determined by elemental analysis was confirmed by ^{13}C NMR spectroscopy and ^{31}P NMR spectroscopy. The existence of phosphorus as phosphate is determined by NMR methods. A protein(s)-inorganic composite is proposed for the molecular structure of Atlantic Cod scale.

Keywords: Fish scale, adsorption, ultimate analysis, SEM, NMR

INTRODUCTION

Over the past few years, biosorption research has focused on the use of readily available material for the accumulation of metal ions (Schiewer, 1999). In particular, the excellent

---

[*] Corresponding author. Email: mustafiz@dal.ca

metal binding capability of marine organisms has been investigated in numerous bio-sorption studies (Holan et al., 1993). Chitosan, an aminopolysaccharide isolated from crab and lobster shells is known to bind metal ions (Yang and Zall, 1984). Recent preliminary studies in our laboratory (Mustafiz, 2002) have indicated that over 90% of heavy metal ions (e.g., lead) are adsorbed by the scales of *Gadus morhua*, one of the 59 species of Atlantic Cod fish. At present, however, there is no evidence to indicate whether Atlantic Cod fish scales are composed of chitin (the insoluble raw material processed to chitosan) or other biopolymers.

The important physical properties relevant to metal ion adsorption by the biomass include porosity and surface area. A high surface area per unit mass is desirable, and the degree of sorption of a metal ion is directly proportional to the number of adsorption sites per unit area. An estimated low micro-pore surface area per unit mass, if compared to a high total surface area, indicates a low porosity of the adsorbent. A low-porosity medium may enhance the retardation factor of the solute and improve the efficiency of its removal. Separation of ionic species by membrane substrates strongly depends on pore size and the charge on the substrate (Tsuru et al., 1991a; Tsuru et al., 1991b; Peeters et al., 1998). A membrane with small pore sizes effectively retains ionic species. But the pore diffusivities of ions diminish with increasing ionic radii of both metal ions and molecular species. The pH of the bulk phase is an important parameter for removing heavy metal ions. In a highly alkaline medium, the substrate is substantially negatively charged due to de-protonation of ionisable groups on its surface. The negatively charged surface has a strong affinity for heavy metal cations. The presence of nucleophilic groups, such as hydroxyl, amino, and carboxylate, is highly advantageous for sorption mechanisms (Dronnet et al., 1997), and results in covalent, hydrogen and electrostatic bonding between the cations in the bulk phase and the nucleophilic sites on the adsorbed phase (Sarwar and Islam, 1997).

In the present study, the applicability of fish scales as a sorbent material for the solid phase extraction of environmental inorganic contaminants, such as lead ions, was investigated, and the degree of heterogeneity of the adsorbent was evaluated using SEM and NMR techniques. The structural analyses provided an understanding of the possible functional groups that are perhaps associated with the adsorbent. Such groups can be correlated to ionic mechanisms that may influence in abatement of the metal ions in hazardous waste streams.

Tests

The original Cod scale was collected from the fisherman's market at Bedford, Nova Scotia, Canada. It was rinsed with distilled water to remove weakly associated sodium ions that may interfere with the over all adsorption phenomenon. A portion of the scales was pulverized to a size of 37-40 micron. To test the adsorption of lead ions, an unpulverized portion was immersed in a solution of lead ions at 1000 ppm for adsorption studies.

To analyze the elements present in the fish scale, an ultimate analysis was conducted at Philip Analytical Services, Bedford. Structural studies of the fish scale by SEM methods were performed at the Department of Metallurgy of Dalhousie University. All NMR experiments were carried out at the Atlantic Region Magnetic Resonance Centre, Department of Chemistry, Dalhousie University, on a Bruker AMX-400 NMR spectrometer. The spectra were obtained using a Bruker 4 mm magic-angle spinning triple-resonance probe with MAS spinning rate of 10 kHz. Both ^{13}C and ^{31}P spectra were obtained using cross-polarization as well as high-power proton decoupling. The contact time in each case was 5 ms and a recycle delay of 10 s was employed. For the ^{31}P spectrum, 32 transients were accumulated, whereas 5512 were accumulated for the ^{13}C spectrum. Carbon chemical shifts were referenced with respect to tetramethylsilane (TMS) at 0 ppm by using adamantane as a secondary reference. This was achieved by setting the highest frequency peak of the adamantane carbon spectrum to 38.6 ppm. Similarly, phosphorus chemical shifts were referenced to 85% H_3PO_4 (aq.) at 0 ppm by using ammonium dihydrogen phosphate (ADP) as a secondary reference. This was achieved by setting the phosphorus signal of ADP to 0.8 ppm.

RESULTS AND DISCUSSION

Structural Analysis

The ultimate analysis results for the composition of fish scale are displayed in Table 1. The amounts of carbon, hydrogen and nitrogen detected are consistent with organic material, and the high level of ash suggests a significant inorganic component. The molar ratio of carbon to nitrogen (3.6:1) in fish scale is remarkably similar to the 3.7:1 ratio calculated from a typical elemental composition for a protein (Lehninger, 1970) and is much less than that given by the formulae for the amino sugar subunits of chitin (8:1) and chitosan (6:1).

Table 1. Elemental Analysis of Atlantic Cod Fish Scale

Analyte	Units (% Wt)	Molar Ratio
Carbon	26.0	1.0
Hydrogen	4.6	2.1
Nitrogen	8.4	0.28
Oxygen	23.2	0.67
Ash	37.8	-

SEM Techniques

In addition to the regular elements of carbon, hydrogen, nitrogen and oxygen as components of fish scale, SEM analysis identified calcium and phosphorus in a ratio of 78 to 22. This ratio is based on the mass of each element.

At 1588X magnification (Figure 1), the non-planar nature of the fish scale is apparent, whereas at a 317X magnification (Figure 2) reveals an extensive 'ridge-like' network. The network consists of 'tooth shaped' features. The 'features' when closely analyzed, confirm that they consist of calcium and phosphorous. An interesting pattern emerges at a 79X magnification (Figure 3). In this view, the substrate of the fish scale, which adheres to the body of the animal, is displayed along with a small portion twisted from the side of the substrate that contacts the aqueous phase. There is no trace of ridges on the body side, yet there is ample distribution of the ridges on its opposite side. Another view at 158X magnification (Figure 4) further demonstrates that the substrate is extensively covered with the 'ridges'. Examination of a specific lobe at 317X magnification shows that the lobe consists of a 'bundle of ropes', which are neatly tied together (Figure 5).

The SEM studies confirmed the presence of lead and arsenic ions on the substrate after adsorption.

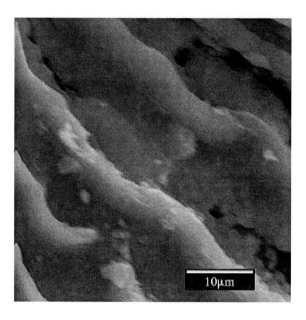

Figure 1. 1588X magnification of substrate (fish scale)

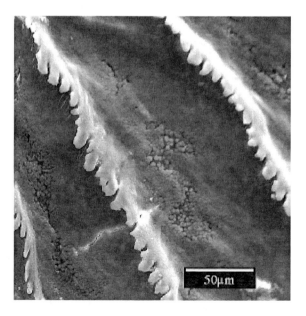

Figure 2. 317X magnification of substrate (fish scale)

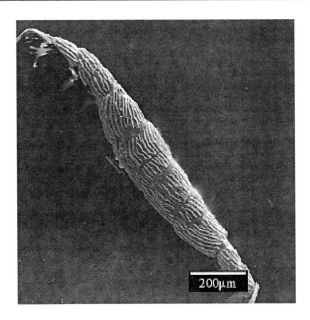

Figure 3. 79X magnification of substrate (fish scale)

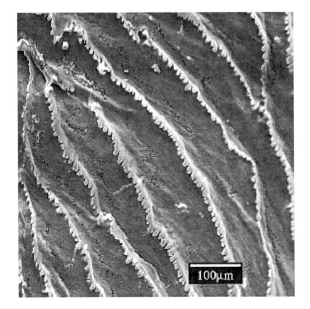

Figure 4. 158X magnification of substrate (fish scale)

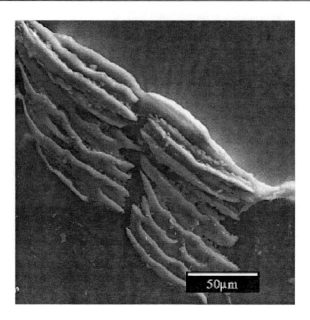

Figure 5. 317X magnification on lobes of the substrate of Figure 2

NMR Analysis

The solid-state ^{13}C NMR spectrum of Cod fish scale is displayed in Figure 6. The spectrum shows prominent resonances for carbonyl carbons (δ 168-180 ppm) and aliphatic carbons (δ 12-66 ppm), as well as less intense resonances around δ 130 and 158 ppm, attributed to aromatic carbons. The ^{31}P NMR spectrum of scale showed only a single resonance at 2.8 ppm.

Polysaccharide is not detected in the sample; no signal characteristic of anomeric carbons is observed near δ 100 ppm in the ^{13}C NMR spectrum. On the other hand, the overall appearance of the spectrum is similar to the spectra of structural proteins, such as collagen (Saitô and Tabeta, 1984) and silk fibroin (Saitô et al., 1984), suggesting protein as the major organic component of fish scale. A general structure of protein is shown in Figure 7. The range of chemical shifts of the signals within the region for aliphatic carbons is consistent with resonances expected for the α-carbons of amino acid residues in the protein, the carbons attached to oxygen in alcohols, and carbons in hydrophobic side chains of amino acids. The presence of amino acids possessing non-polar side chains (alkyl) is consistent with the adsorbent's low solubility in water. The possible presence of carboxylic acids and primary amines as functional groups (at side ends of the molecule) in the structure of the adsorbent are displayed in Figure 8.

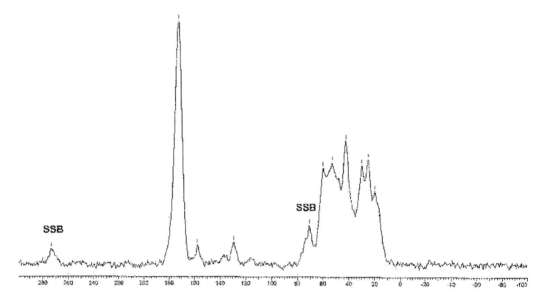

Figure 6. ^{13}C CP/MAS spectrum of the pulverized Cod scale sample. The peaks denoted by "SSB" are spinning sidebands. The spectrum represents the accumulation of 5512 scans

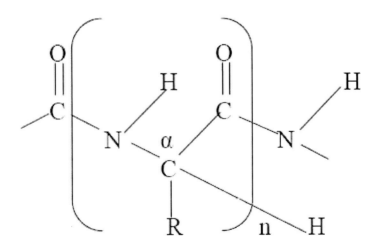

Figure 7. A general structure of protein showing the α-carbon and R for the various side-chain structures of the amino acid residues

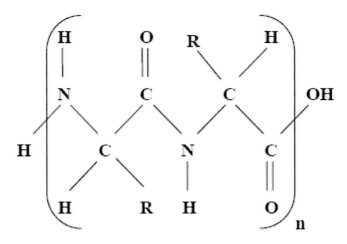

Figure 8. Proposed molecular structure of the adsorbent with primary amine and carboxyl groups at the terminals of the molecule

The chemical shift of the ^{13}C resonance assigned to carbonyl groups is very similar to that observed in other proteins (Saitô and Tabeta, 1984; Saitô et al., 1984) and therefore corresponds to amide carbonyls in the protein backbone and carboxyl groups in the side chains of aspartic and glutamic acid residues. The very high intensity of the carbonyl resonance, however, suggests the presence of another species containing carbon that resonates at a similar chemical shift. When the fish scale was treated with dilute aqueous HCl, an approximately 20% loss of mass was observed, suggesting the conversion of carbonate ion to carbon dioxide. Carbonate ion could be associated with the calcium detected by SEM and contribute to the intensity of the carbonyl resonance in the ^{13}C-NMR spectrum. For example, the carbon chemical shift of solid calcium carbonate (limestone) has been measured by Skála and Rohovec (1998) to range between 166 and 169 ppm, consistent with this interpretation.

The phosphorus indicated by SEM was confirmed by a strong resonance in the ^{31}P NMR spectrum with a chemical shift of 2.8 ppm (Figure 9). The absence of spinning sidebands in the spectrum also indicates a small chemical shielding anisotropy associated with this species. These results are consistent with the presence of orthophosphate in the sample and furthermore show a striking similarity with the ^{31}P NMR spectrum obtained in the study by Turner et al. (1986) for calcium phosphate. These workers measured phosphorus chemical shifts and anisotropies in a series of orthophosphates using a number of different cations. For the calcium cation, they measured an isotropic shift of +3.0 ppm and a negligibly small anisotropy, in very close agreement with what we have observed for the fish scale sample.

Collectively, the NMR, SEM and ultimate analysis data collected in this study suggest that the Cod fish scale is a composite consisting of protein, calcium carbonate, and calcium phosphate. Like chitin, which also possesses amide groups (Rhee et al., 1998), strong hydrogen bonding between water molecules and nitrogen sites within the biopolymer is a main factor behind swelling of fish scales.

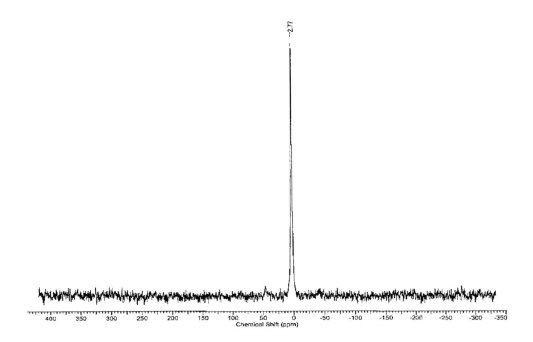

Figure 9. ^{31}P NMR spectrum of the pulverized Cod scale sample. The peak denotes the existence of phosphate by NMR method

CONCLUSION

The SEM analysis illustrates the presence of significant amount of calcium and phosphorus in the scale material. Primarily the experimental results have also shown that the elements calcium and phosphorus exist as both calcium carbonate and calcium phosphate. The NMR technique confirms the heterogeneity and long chain nature of the adsorbent. The substrate of the fish scale is composed of carbonyl, amide, alkyl and other functional groups.

ACKNOWLEDGMENTS

The authors would like to acknowledge the financial assistance provided by NSERC and ACPI for conducting this research. We also acknowledge the assistance of the Department of Chemistry, Dalhousie University in allowing us access to their laboratory facilities. Mr. Mustafiz gratefully acknowledges the financial contribution of the Killam Foundation.

REFERENCES

Dronnet, V.M., Renard, C.M.G., Axelos, M.A.V., Thibault, J.F., 1997. Adsorption techniques. *Carbohydrate Polymers*. 34, 73-82.

Holan, Z.R., Volesky, B., Prasteyo, I., 1993. Biosorption of cadmium by biomass of marine algae. *Biotechnology Progress.* 41, 819-825.

Lehninger, A.L., 1970. Biochemistry: The molecular basis of cell structure and function. Worth Publishers, New York, pp. 55-56.

Mustafiz, S., 2002. A novel method for heavy metal removal from aqueous streams. *MASc. Thesis.* Dalhousie University, Canada.

Peeters, J.M.M., Boom, J.P., Mulder, M.H.V, Strathmann, H., 1998. Retention measurements of nano filtration membranes with electrolyte solutions. *Journal of Membrane Science.* 145, 199-203.

Rhee, J., Jung, M.W., Paeng, K., 1998. Evaluation of chitin and chitosan as a sorbent for preconcentration of phenol and chlorophenols in water, *Analytical Sciences.* 14, December.

Saitô, H., Tabeta, R., 1984. A high-resolution ^{13}C-NMR study of collagenlike and collagen fibrils in solid state studied by the cross-polarization-magic angle-spinning method. Manifestation of conformational-dependent ^{13}C chemical shifts and application to conformational characterization. *Biopolymers.* 23, 2279-2297.

Saitô, H., Tabeta, R., Asakura, T., Iwanaga, Y., Shoji, A., Ozaki, T., Ando, I., 1984. High-resolution ^{13}C NMR study of silk fibroin in the solid state by the cross-polarization-magic angle spinning method, conformational characterization of silk I and silk II type forms of Bombyx mori fibroin by the conformation-dependent ^{13}C chemical shifts. *Macromolecules.* 17, 1405-1412.

Sarwar, M., Islam, M.R., 1997. A non Fickian surface excess model for chemical transport through fractured porous media. *Chemical Engineering Communications.* 160, 1-34.

Schiewer, S., 1999. Modelling complexation and electrostatic attraction in heavy metal biosorption by sargassum biomass. *Journal of Applied Phycology.* 11, 79-87.

Skála, R., Rohovec, J., 1998. Magic angle spinning nuclear magnetic resonance spectroscopy of shocked limestones from the Steinheim Crater. Proc. 61st meteoritical society meeting. Dublin, Ireland.

Tsuru, T., Nakao, S., Kimura, S., 1991a. Calculations of ion rejection by extended Nernst-Planck equation with charged reverse osmosis membranes for single and mixed electrolyte solutions. *Journal of Chemical Engineering.* Japan, 24, 511-117.

Tsuru, T., Urairi, M., Nakao, S., Kimura, S., 1991b. Reverse osmosis of single and mixed electrolytes with charged membranes: Experiments and Analysis. *Journal of Chemical Engineering.* Japan, 24, 518-523.

Turner, G.L., Smith, K.A., Kirkpatrick, J., Oldfield, E., 1986. Structure and cation effects on phosphorus-31 NMR chemical shifts and chemical-shift anisotropies of orthophosphates. *Journal of Magnetic Resonance.* 70, 408-415.

Yang, T.C., Zall, R.R., 1984. Adsorption of metals by natural polymers generated from seafood processing wastes. *Industrial Engineering Chemistry.* 23, 168-172.

In: Nature Science and Sustainable Technology
Editor: M. R. Islam, pp. 83-118

ISBN: 978-1-60456-009-1
© 2008 Nova Science Publishers, Inc.

Chapter 5

REVERSING GLOBAL WARMING

A. B. Chhetri and M. R. Islam
Department of Civil and Resource Engineering
Dalhousie University, Halifax, Canada

ABSTRACT

Global warming has been a subject of discussion from late seventies. It is perpetrated that the building up of carbon dioxide in the atmosphere results in irreversible climate change. Even though carbon dioxide has been blamed as the sole cause for the global warming, there is no scientific evidence that all carbon dioxides are responsible for global warming. A new theory has been developed, which shows that all carbon dioxides do not contribute to global warming. For the first time, carbon dioxide is characterized based on various criteria, such as the origin, the pathway it travels, and the isotope number. In this paper, the current status of greenhouse gas emissions from various anthropogenic activities is summarized. Role of water in global warming has been discussed. Various energy sources are classified based on their global efficiencies. The assumptions and implementation mechanisms of the Kyoto Protocol have been critically reviewed. It is argued that Clean Development Mechanism of the Kyoto Protocol has become the 'license to pollute' due to its improper implementation mechanism. The conventional climatic models are deconstructed and guidelines for new models are proposed in order to achieve the true sustainability in the long term. A series of sustainable technologies that produce natural CO_2, which do not contribute to global warming has been presented. Various zero-waste technologies that have no negative impact on the environment are keys to reverse the global warming. Because synthetic chemicals, which are inherent to the current technology development mode, are primarily responsible for global warming, there is no hope for reversing global warming without fundamental changes in technology development. The new technology development mode must foster the development of natural products, which are inherently beneficial to the environment.

Keywords: Global warming, climate change, sustainable technologies

INTRODUCTION

Global warming has been a subject of discussion from late 1970s. Indeed the discussion of the possibility that a build-up of carbon dioxide in the atmosphere results in irreversible climate change has been transformed into a "controversy" of the type seen all too often on every other subject: a "pro" *versus* "con" proposition is advanced, dividing people according to their support for one side or the other, all before anything objective and scientific in connection with the originating subject-matter is even established.

As far as the science of the question goes: despite various international and government organizations have set series of standards to reduce the carbon dioxide level in the atmosphere due to anthropogenic activities, the current climatic models show that the global temperature is still increasing. Carbon dioxide has been blamed as the sole cause for the global warming, even though there is no scientific evidence that all carbon dioxides are responsible for global warming. Precisely to address this critical gap, a detailed analysis of greenhouse gas emission starting from pre-industrial era, industrial age to the (for some) "golden era" of petroleum has been carried out in this paper.

A very large amount of pseudo-science is already afoot on all aspects of this question, much of it used to divide – if not indeed aimed in the first place at dividing – public opinion over whether Nature or Humanity is the chief culprit. This state of affairs has opened the door to proposing all manner of band-aid solutions that share the trait of enabling peoples of the global North to continue dreaming of two SUVs for every garage while peoples of the South can continue to fantasize about motorizing their bicycle or oxcart – in other words: the *status quo*, with one foot on the accelerator while the argument carries on as to when to apply the brake. The crying need for the serious scientific approach taken in the present work has never been greater. On this question, paraphrasing Albert Einstein, it can truly be said that the system that got us into the problem is not going to get us out. Absent a comprehensive characterization of CO_2 in all its possible roles and forms as a starting-point, any attempt to analyze the tangle of symptoms identified with global warming, or to design any solution, based on univariate correlations or even correlations of multiple variables, but assuming that the effects of each variable can be superposed linearly and still mean anything, must collapse under the weight of its very incoherence. The absurdity is so well known that one popular graph on the Internet depicts a strictly proportional increase in incidences of piracy in all the world's oceans as a function of increasing global temperature.

The current status of greenhouse gas emissions due to industrial activities, automobile emissions, biogenic and natural sources is systematically presented here. In this paper, a newly developed theory has been detailed: all carbon dioxides are not same. Thus, not all carbon dioxides may be contributing to global warming. For the first time, carbon dioxide is characterized based on such normally ignored criteria such as its origin, the pathway it travels, isotope number and age of the fuel source from which it was emitted. Fossil fuel exploration, production, processing and consumption are major sources of carbon dioxide emissions; here various energy sources are characterized based on their efficiency, environmental impact and quality of energy based on the new criteria. Different energy sources follow different paths from origin to end-use and contribute the emissions differently.

A detailed analysis has been carried out on potential precursors to global warming. The focus is on supplying a scientific basis as well as practical solutions identifying the roots of the problem. Shortcomings in the conventional models have been identified based on this evaluation.

The sustainability of conventional global warming models has been argued. Here, these models are deconstructed and guidelines for new models based on new sustainability criteria. Conventional energy production and processing uses various toxic chemicals and catalysts that are very harmful to the environment. Moreover, all energy systems are totally dependent on fossil fuel at least as the primary energy input or in the form of embodied energy. This paper offers unique solutions to overcome such problems, based on truly green technologies that satisfy the new sustainability criteria. These green energy technologies are highly efficient technologies producing with zero net waste.

In this paper, various energy technologies are ranked based on their global efficiency. For the first time, this research offers energy development techniques that produce what might best be described as "good or natural CO_2" which do not contribute to global warming. A thorough discussion of natural transport phenomena, specifically the role of water and its interaction with various energy sources and climate change taking into account the memory of water, is also undertaken in this work. Conventional models are evaluated based on the long-term impact of CO_2 and their contribution to global warming. It is concluded that conventional energy development systems and global warming models are based on ignorance. Only knowledge-based technology development offers solutions to the global warming.

HISTORICAL DEVELOPMENT

The history of technological development from the pre-industrial age to petroleum era has been reviewed. There is a colloquial expression to the effect that exact change plus faith in the Almighty will always get you downtown on the public transit service. On the one hand, with or without faith, all kinds of things could happen with the public transit service, before the matter of exact fare even enters the picture. On the other hand, with or without exact fare, other developments could intervene to alter the availability of the service and even cancel it. This helps isolate one of the key difficulties in uncovering and elaborating the actual science of increased carbon-dioxide concentrations in the atmosphere. All kinds of activities can increase CO_2 output into the atmosphere; but precisely which activities can be held responsible for consequent global warming or other deleterious impacts? Both the activity and its CO_2 output are necessary, but neither by itself is sufficient, for establishing what the impact may be and whether it is deleterious.

Pre-Industrial

One commonly encountered argument attempts to frame the historical dimension of the problem more or less as follows: once upon a time, the scale of Humanity's efforts at securing a livelihood was insufficient to affect overall atmospheric levels of CO_2. The implication is that, with the passage of time and the development of ever more extensive technological intervention in the natural-physical environment by Humanity, everything just got worse. In a contemporary world that has systematically removed ever further from the human person any living connection with the gathering or application of meaningful knowledge, such typically linearized evolutionary analyses may pass for meaningful exegesis. However, from prehistoric times onward, there have been important periods of climate change whose causes could not have had anything to do with human intervention in the environment on anything approaching the scale that is blamed widely today for "global warming". Nevertheless, these had consequences that were extremely significant and even devastating for wide swaths of subsequent human life on this planet.

One of the best-known was the period of almost two centuries of cooling in the northern hemisphere during the 13th and 14th centuries CE, in which Greenland is said to have acquired much of its most recent ice cover. This definitively brought to an end any further attempts at colonising the north and northwest Atlantic by Scandinavian tribes (descended from the Vikings), creating the opening for later commercial fisheries expansion into the northwest Atlantic by Basque, Spanish, Portuguese and eventually French and British fishermen and fishing enterprises – the starting-point of European colonization of the North American continent.

Industrial Age

Even one-off events like the volcanic eruption in the Indonesian archipelago in 1816, which spewed an enormous volume of dust into the atmosphere traveling around the globe in the jet stream and led to the "year with no summer" in Europe and the northern half of North America, incurred tremendous consequences. In 1817, grain crops on the continent of Europe failed. In industrial Great Britain, where the factory owners and their politicians boasted how that country's relative (compared to the rest of the world) highly advanced industrial economy had overcome the "capriciousness of Nature", hunger and famine actually stalked the English countryside for the first time in more than a century and a half. The famine conditions were blamed on the difficulties attending the import of extra supplies of food from the European continent, and led directly to a tremendous and unprecedented pressure to eliminate the Corn Laws – the system of high tariffs protecting English farmers and landlords from the competition of cheaper foodstuffs from Europe or the Americas. Politically, the industry lobby condemned the Corn Laws as the main obstacle to cheap food, winning broad public sympathy and support. Economically, the Corn Laws actually operated to keep hundreds of thousands employed in the countryside on thousands of small agricultural plots, at a time when the demands of expanding industry required uprooting and forcing this rural population to work as factory labourers. Increasing the industrial reserve army would enable British industry to reduce wages. Capturing command of that new source of ever cheaper labour was in fact the industrialists' underlying aim.

Without the famine of "the year with no summer", it seems unlikely British industry would have hit upon the political device of targeting the Corn Laws for elimination as the road on which to blast its way into dominating world markets. Even then, because of the still prominent involvement of the anti-industrial lobby of aristocratic landlords who dominated the House of Lords, it would take British industry another nearly 30 years, but between 1846 and 1848 Parliament eliminated the Corn Laws, industry captured access to a desperate workforce fleeing the ruin brought to the countryside, and overall industrial wages were driven sharply downwards. On this train of economic development, the greatly increased profitability of British industry took the form of a vastly whetted appetite for new markets at home and abroad, including the export of important industrial infrastructure investments in "British North America", *i.e.*, Canada; Latin America; and India. Extracting minerals and other valuable raw materials for processing into new commodities in this manner brought an unpredictable level of further acceleration to the industrialisation of the globe in regions where industrial capital had not accumulated significantly either because traditional development blocked its role or because European settlement remained sparse.

Age of Petroleum

The world economy entered the Age of Petroleum mostly since the rise of industrial-financial monopoly in one sector of production after another in Europe and America, before and following the First World War. Corresponding to this has been the widest possible extension of chemical engineering – especially the chemistry of hydrocarbon combination, hydrocarbon catalysis, hydrocarbon manipulation and rebonding – on which the refining and processing of crude oil into fuel and myriad byproducts such as plastics and other synthetic materials crucially depend. As a result, there is today no activity, be it production or consumption, in any society that is tied to the production and distribution of such output in which adding to the CO_2 burden in the atmosphere can be avoided or significantly mitigated.

In these developments, carbon and CO_2 are in fact vectors carrying many other actually toxic compounds and byproducts of these chemically-engineered processes. Atmospheric absorption of carbon and CO_2 from human activities or other natural non-industrial activities would normally be continuous. However, what occurs with hydrocarbon complexes combined with inorganic and other substances that occur nowhere in nature is much less predictable, and – on the available evidence – not benign, either. From a certain standpoint, there is a logic in attempting to estimate the effects of these other phenomena by taking carbon and CO_2 levels as vectors. However, there has never been any justification to assume the CO_2 level itself is the malign element.

Such a notion is a non-starter as science in any event – which raises the even sharper question: just what does science have to do with it? There is today no large petrochemical company or syndicate that has not funded some study, group, or studies or groups, interested in CO_2 levels as a global warming index – whether to discredit or to affirm such a connection. It is difficult to avoid the obvious inference that these very large enterprises, fiercely competing to retain their market shares against rivals, do not have a significant stake in engineering a large and permanent split in public opinion based on confusing their intoxication of the atmosphere with rising CO_2 levels. Whether the consideration is refining for automobile fuels, processing synthetic plastics, or concocting synthetic crude, behind a great deal of the propaganda about "global warming" stands a huge battle among oligopolies, cartels and monopolies over market share. The science of "global warming" is precisely the only route on which to separate the key question of what is necessary to produce goods and services that are Nature-friendly from the toxification of the environment as a byproducts of the anti-Nature bias of chemical engineering in the clutches of the oil barons.

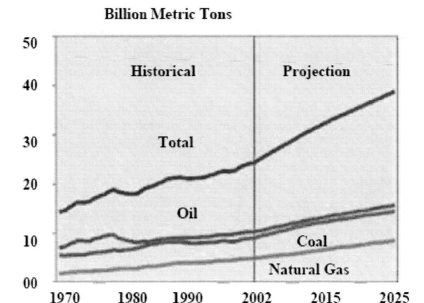

Figure 1. World CO_2 Emissions by oil, coal and natural gas, 1970-2025 (adopted from EIA, 2004)

CURRENT STATUS OF GREENHOUSE GAS EMISSION

Current status of greenhouse gas emissions from various anthropogenic activities is summarized. Industrial activities, especially related to the burning of fossil fuels, are major contributors of global greenhouse gas emissions. Climate change due to anthropogenic greenhouse gas (GHG) emissions is a growing concern for global society. In the third assessment report, the Intergovernmental Panel on Climate Change (IPCC) provides the strongest evidence so far that the global warming of the last 50 years is due largely to human activity and the CO_2 emissions that arise when burning fossil fuel (Farahani et al., 2004).

It has been reported that the CO_2 level now is at the highest point in 125,000 years (Service, 2005). Approximately 30 billion tons of CO_2 is released from fossil fuel burning each year (Figure 1). The CO_2 concentration level in the atmosphere traced back in 1750 was reported to be 280±10ppm (IPCC, 2001a). It has risen continuously since then and the CO_2 level reported in 1999 was 367 ppm. The present atmospheric CO_2 concentration level has not been exceeded during the past 420,000 years (IPCC, 2001b; Houghton et al., 2001; Houghton, 2004).

The latest 150 years were a period of global warming (Figure 2). Global mean surface temperatures have increased 0.5-1.0°F since the late 19th century. The 20th century's 10 warmest years all occurred in the last 15 years of the century. Of these, 1998 was the warmest year on record. Sea level has risen 4-8 inches globally over the past century. Worldwide precipitation over land has increased by about one percent.

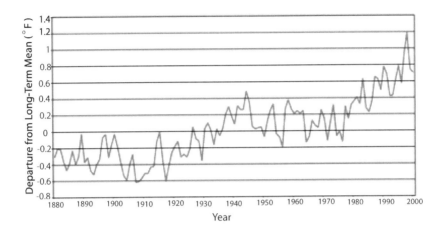

Figure 2. Global Temperature Changes from 1880 to 2000 (Modified after EPA Global Warming site: US National Climate Data Center 2001)

The industrial emission of CO_2 consists of process emission and production emission. Coal mining, oil refining, gas processing, petroleum fuel combustion, pulp and paper industry, ammonia, petroleum refining, iron and steel, aluminum, electricity generation and cement production are the major industries responsible for producing various types of greenhouse gases. Besides these industrial sources, transportation sector has also large share of greenhouse gas emission. Greenhouse gas emission from bioresources is also significant. However, National Energy Board of Canada does not consider CO_2 from biomass as contribution to greenhouse problems (Hughes and Scott, 1997). The justification emerges from the fact that greenhouse gas emission from bioresources such as fuelwood, agricultural waste, charcoal is carbon neutral as the plants synthesize this CO_2. However, if various additives are added during the production of fuel such as pellet making and charcoal production, the CO_2 produced is no longer carbon neutral. For instance, pellet making involves the addition of binders such as carbonic additives, coal and coke breeze which emit carcinogenic benzene as a major aromatic compound (Chhetri at al., 2006a). The CO_2 contaminated with such chemical additives is not favored by plants for photosynthesis and as a result, CO_2 will be accumulated in the atmosphere. Moreover, deforestation, especially the unsustainable harvesting of biomass due to urbanization and to fulfill the industrial biomass requirement also results in net CO_2 emission from bioresources.

The worldwide CO_2 emission from the consumption of fossil fuels was 24,409 million metric tons in 2002 and it is projected to reach to 33284 million metric tons in 2015 and 38,790 million tons in 2025 (IEO, 2005). The worldwide CO_2 production from consumption and flaring of fossil fuel in 2003 was 25,162.07 million metric tones. The USA alone had a share of 5802.08 million tones of CO_2 emission in 2003 (IEA, 2005). Current CO_2 emission levels are expected to continue increasing in the future as fossil fuel consumption is sharply increasing (WEC, 2006). The projection showed that emission from all sources are emitted to grow by 36% in 2010 (to 18.24 Gt/y) and by 76% in 2020 to 23.31 Gt/y (compared to the 2000 base level). Variation of CO_2 concentration at different time scales is presented in Figure 3 This figure shows the increase in CO_2 emission exponentially after 1950. However, present methodology does not classify CO_2 based on its source. Industrial activities during this period also went up exponentially. Because of this industrial growth and extensive use of fossil fuels, the level of 'industrial' CO_2 emission increased sharply (Figure 4). The worldwide supply in 1970 was approximately 49 million barrels per day but the supply increased to approximately 84 million barrels per day (EIA, 2006). At the same time, the level of 'natural' CO_2 which comes by burning biomass went down due to deforestation. However, researchers, industry and government are focused on the total CO_2, which is not correct in terms of its impacts on global warming. NOAA (2005a) defined annual mean growth rate of CO_2 as the sum of all CO_2 added to, and removed from, the atmosphere during the year by human activities and by natural processes. Natural CO_2 cannot be same as that of industrial CO_2 and should be examined separately.

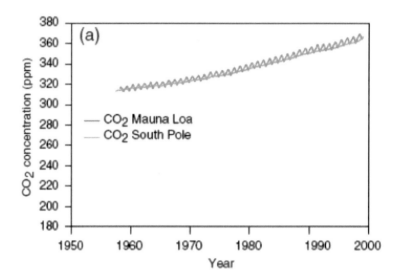

Figure 3. Variation in atmospheric CO_2 concentration (IPCC, 2001a)

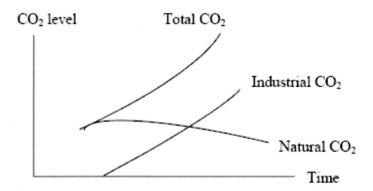

Figure 4. Total, industrial and natural CO_2 trend

Some recent studies reported that the human contribution to global warming is negligible (Khilyuk and Chilingar, 2003; Khilyuk and Chilingar, 2004). The global forces of nature such as solar radiation, outgassing from the Ocean and the atmosphere and microbial functions are driving the Earth's climate (Khilyuk and Chilingar, 2006). These studies showed that the CO_2 emissions from human induced activities is far less in quantity than the natural CO_2 emission from Ocean and volcanic eruptions. This line of argument is used by others to demonstrate that the cause of global warming is at least a contentious issue (Goldschmidt, 2005). These studies fail to explain the differences between the natural and human induced CO_2 and their impacts on global warming. Moreover, the CO_2 from ocean and natural forest fires were a part of the natural climatic cycle even when no global warming was noticed. All the global forces mentioned by (Khilyuk and Chilingar, 2006) are also affected by human interventions. For example, more than 85,000 chemicals being used worldwide for various industrial and agricultural activities are exposed in one or the other way to the atmosphere or ocean water bodies that contaminate the CO_2 (Global Pesticide Campaigner, 2004) The CO_2 produced from fossil fuel burning is not acceptable to the plants for their photosynthesis and for this reason, most organic plant matters are depleted in carbon ratio $\delta^{13}C$ (Farquhar et al., 1989; NOAA, 2005b). Finally the notion of 'insignificant' has been used in the past to allow unsustainable practices, such as pollution of harbors, commercial fishing and massive production of toxic chemicals that were deemed to be "Magic Solutions" (Khan and Islam, 2005a; Khan and Islam, 2006a). Today, banning of chemicals and pharmaceutical products has become almost a daily affair (The Globe and Mail, 2006; The New York Times, 2006). None of these products was deemed 'significant' or harmful when they were introduced. Khan and Islam (2006b) have recently catalogued an array of such ill-fated products that were made available to 'solve' a critical solution (Environment Canada, 2006). In all these engineering observations, a general misconception is perpetrated, that is: if the harmful effect of a product can be tolerated in the short-term, the negative impact of the product is 'insignificant'.

Khilyuk and Chilingar (2006) explained an adiabatic model developed by Khilyuk et al. (1994) for the atmosphere together with a sensitivity analysis to evaluate the effects of human induced CO_2 emissions on the global temperature. The model showed that due to the human induced CO_2, the global temperature rise is negligible. However, considering that the adiabatic condition in the atmosphere is one of the most linear thoughts and the basic assumption of this model is incorrect (Khan et al., 2006a). For an adiabatic condition, the following three assumptions are made: perfect vacuum between system and surrounding area, perfect reflector around the system like the thermo flux mechanism to resist radiation and zero heat diffusivity material that isolates the system. None of these conditions can be fulfilled in the atmosphere. Moreover, the study reported that increased emissions of carbon dioxide and water vapour are important for agriculture and biological protection and the CO_2 from fossil fuel combustion is non toxic. However, their finding is in contradiction to the fact that plants discriminate against heavier CO_2 and favor CO_2 with lighter carbon isotope ratios. As all chemicals are not the same, all CO_2 are not the same. The CO_2 from power plants is highly toxic as various toxic chemicals are added during the refining and processing of fossil fuels (Chhetri et al., 2006a; Chhetri and Islam, 2006a; Chhetri and Islam, 2006b; Khan and Islam, 2006b). Since the CO_2 from fossil fuel burning is contaminated with various toxic chemicals, plants do not readily synthesize it. Note that practically all catalysts used are either chemically synthesized or are denatured by concentrating them to a more beneficial state (Khan and Islam, 2006b). The CO_2 rejected from plants accumulates in the atmosphere and is fully responsible for global warming. According to Thomas and Nowak (2006), human activities have already demonstrably changed global climate, and further, much greater changes are expected throughout this century. The emissions of CO_2 and other greenhouse gases will further accelerate global warming. Some future climatic consequences of human induced CO_2 emissions, for example some warming and sea-level rise, cannot be prevented, and human societies will have to adapt to these changes. Other consequences can perhaps be prevented by reducing CO_2 emissions.

Figure 5 is the crude oil pathway. The crude oil is refined to convert into various products including plastics. More than four million tons of plastics are produced from 84 million barrels of oil per day. It has been further reported that plastic burning produces more than 4000 toxic chemicals, 80 of which are known carcinogens (Islam, 2004).

Crude oil →Gasolene+Solid residue+diesel+kerosene+volatile HC+ numerous petroleum products

Solid residue +hydrogen+metal (and others) →plastic

Figure 5. The crude oil pathway (Islam, 2004)

In addition to the CO_2, various other greenhouse gases have contributed to global warming. The concentration of other greenhouse gases has increased significantly in the period between 1750-2001. Several classes of halogenated compounds such as chlorine, bromine, fluorine are also greenhouse gases and are the direct result of industrial activities. None of these compounds was in existence before 1750 but are found in significant concentration in the atmosphere after that period (Table 1). Chlorofluorocarbons (CFCs), hydrohloroflorocarbons (HCFCs) which contains chlorine and halocarbons such as bromoflorocarbons which contain bromine are considered potent greenhouse gases. The sulfur hexafluoride (SF_6) which is emitted from various industrial activities such as aluminum industry, semi-conductor manufacturing, electric power transmission and distribution, magnesium casting and from nuclear power generating plants also considered a potent greenhouse gas. Table 2 shows that the concentration of these chemicals has significantly increased in the atmosphere after 1750. For example, the CFC-11 was not present in the atmosphere before 1750. However, the concentration after 1750 reached to 256 ppt after 1750. It is important to note here that these chemicals are totally synthetic in nature cannot be manufactured under natural conditions. This would explain why the future pathway of these chemicals is so rarely reported.

The transportation sector consumes a quarter of the world's energy, and accounts for some 25% of total CO_2 emissions, 80% of which is attributed to road transport (EIA, 2006). Projections for Annex I countries indicate that, without new CO_2 mitigation measures, road transport CO_2 emissions might grow from 2500 million tones in 1990 to 3500 to 5100 million tones in 2020. The fossil fuel consumption by the transportation sector is also sharply increasing in the Non Annex I countries as well. Thus the total greenhouse gas emission from transportation will rise in the future. It is reported that as much as 90% of global biomass burning is human-initiated and that such burning is increasing with time (NASA, 1999). Forest products are the major source of biomass along with agricultural as well as household wastes. The CO_2 from biomass has long been considered to be the source of feedstock during photosynthesis by plants. Therefore the increase in CO_2 from biomass burning can not be considered to be unsustainable, as long as the biomass is not contaminated through 'processing' before burning. CO_2 from unaltered biomass is distinguished from CO_2 emitted from processed fuels. To date, any processing involves the addition of toxic chemicals. Even if the produced gases do not show detectable concentration of toxic products, it is conceivable that the associated CO_2 will be different from CO_2 of organic origin. The CO_2 emission from biomass which is contaminated with various chemical additives during processing has been calculated and deducted from the CO_2 which is good for the photosynthesis that does not contribute to global warming.

Table 1. Concentrations, global warming Potentials (GWPs), and atmospheric lifetimes of GHGs

Gas	Pre-1750 concentration	Current topospheric concentration	GWP (100-yr time horizon)	Life time (years)
carbon dioxide (CO_2)	280 ppm	374.9	1	varies
methane (CH_4)	730ppb	1852ppb	23	12
nitrous oxide (N_2O)	270	319ppb	296	114
CFC-11 (trichlorofluoromethane) (CCl_3F)	0	256 ppt	4600	45
CFC-12 (dichlorodifluoromethane) (CCl_2F_2)	0	546 ppt	10600	100
CFC-113 (trichlorotrifluoroethane) ($C_2Cl_3F_3$)	0	80 ppt	6000	85
carbon tetrachloride (CCl_4)	0	94 ppt	1800	35
methyl chloroform (CH_3CCl_3)	0	28 ppt	140	4.8
HCFC-22 (chlorodifluoromethane) ($CHClF_2$)	0	158 ppt	1700	11.9
HFC-23 (fluoroform) (CHF_3)	0	14 ppt	12000	260
perfluoroethane (C_2F_6)	0	3 ppt	11900	10000
sulfur hexafluoride (SF_6)	0	5.21 ppt	22200	3200
trifluoromethyl sulfur pentafluoride (SF_5CF_3)	0	0.12 ppt	18000	3200

Source: IPCC, 2001a

CLASSIFICATION OF CARBONDIOXIDE

Carbon dioxide is considered to be the major precursor for current global warming problems. Previous theories were based on the "Chemicals are Chemicals" approach of the two time Nobel Laureate Linus Pauling's vitamin C and antioxidant experiments. This approach advanced the principle that whether it is from natural or synthetic sources and irrespective of the pathways it travels, vitamin C is same. This approach essentially disconnects a chemical product from its historical pathway. Eventhough, the role of pathways has been understood by many civilizations for centuries, systematic studies questioning their principle is a very recent phenomenon. For instance, it was only recently reported that vitamin C did not lower death rates among elderly people, and may actually have increased the risks of dying (Gale et al., 1995). Moreover, ß carotene supplementation may do more harm than good in patients with long cancer (Josefson, 2003). Obviously such a conclusion can not be made if subjects were taking vitamin C from natural sources. In fact, the practices of people who live the longest lives indicate clearly that natural products do not have any negative impact on human health (York, 2003). More recently it is reported that antioxidant supplement including vitamin C should be avoided by patients being treated as the cancer cells gobbles up vitamin C faster than normal cells which might give greater protection for tumors rather than normal cells (Agus et al.,1999). Antioxidants that are present in nature are known to act as anti-aging agents. Obviously these antioxidants are not the same as those synthetically manufactured. The previously used hypothesis that "Chemicals are Chemicals" fails to distinguish between the characteristics of synthetic and natural vitamins and antioxidants. The impact of synthetic antioxidants and vitamin C in body metabolism would be different than that of natural sources. Numerous other cases can be cited demonstrating that the pathway involved in producing the final product is of utmost importance. Some examples have recently been investigated by Islam and co-workers (Islam, 2004; Khan et al., 2006b; Khan and Islam, 2006b; Zatzman and Islam, 2006). If the pathway is considered, it becomes clear that organic produce is not the same as non organic produce, natural products are not the same as bioengineered products, natural pesticides are not the same as chemical pesticides, natural leather is not same as synthetic plastic, natural fibers are not the same as synthetic fibers, natural wood is not same as fiber- reinforced plastic etc. (Islam, 2006). In addition to being the only ones that are good for the long-term, natural products are also extremely efficient and economically attractive. Numerous examples are given in Khan and Islam (2006b). Unlike synthetic hydrocarbons, natural vegetable oils are reported to be easily degraded by bacteria (AlDarbi et al., 2005). Application of wood ash to remove arsenic from aqueous streams is more effective than removing by any synthetic chemicals (Rahman, et al., 2004; Wassiuddin et al., 2002). Using the same analogy, carbon dioxide has also been classified based on the source from where it is emitted, the pathway it traveled and age of the source from which it came from (Chhetri et al., 2006a;Khan and Islam, 2006b).

Carbon dioxide is classified based on a newly developed theory. It has been reported that plants favor a lighter form of carbon dioxide for photosynthesis and discriminate against heavier isotopes of carbon (Farquhar et al., 1989). Since the fossil fuel refining involves the use of various toxic additives, the carbon dioxide emitted from these fuels is contaminated and is not favored by plants. If the CO_2 comes from wood burning, which has no chemical additives, this CO_2 will be most favored by plants. This is because the pathway the fuel travels from refinery to combustion devices makes the refined product inherently toxic (Chhetri et al., 2006a). The CO_2 which the plants do not synthesize accumulates in the atmosphere. The accumulation of this rejected CO_2 must be accounted in order to assess the impact of human activities on global warming. This analysis provided a basis for discerning natural CO_2 from "man-made CO_2", which could be correlated with global warming.

ROLE OF WATER IN GLOBAL WARMING

Flow of water in different forms has a great role in climate change. Water is one of the components of natural transport phenomenon. Natural transport phenomenon is a flow of complex physical processes. The flow process consists of production, storage and transport of fluids, electricity, heat and momentum (Figure 5). The most essential material components of these processes are water and air which are also the indicators of natural climate. Oceans, rivers and lakes form both the source and sink of major water transport systems. Because water is the most abundant matter on earth, any impact on the overall mass balance of water is certain to impact the global climate. The interaction between water and air in order to sustain life on this planet is a testimony to the harmony of nature. Water is the most potent solvent and also has very high heat storage capacity. Any movement of water through the surface and the Earth's crust can act as a vehicle for energy distribution. However, the only source of energy is the sun and sunlight is the most essential ingredient for sustaining life on earth. The overall process in nature is inherently sustainable, yet truly dynamic. There isn't one phenomenon that can be characterized as cyclic. Only recently, scientists have discovered that water has memory. Each phenomenon in nature occurs due to some driving force such as pressure for fluid flow, electrical potential for the flow of electricity, thermal gradient for heat, and chemical potential for a chemical reaction to take place. Natural transport phenomena cannot be explained by simple mechanistic views of physical processes by a function of one variable. Even though Einstein pointed out the possibility of the existence of a fourth dimension a century ago, the notion of extending this dimensionality to infinite numbers of variables is only now coming to light (Islam, 2006a). A simple flow model of natural transport phenomenon is presented in Figure 5. This model shows that nature has numerous interconnected processes such as production of heat, vapour, electricity and light, storage of heat and fluid and flow of heat as well as fluids. All these processes continue for infinite time and are inherently sustainable. Any technologies that are based on natural principles are sustainable (Khan et al., 2005a; Khan and Islam, 2006b).

Water plays a crucial role in the natural climatic system. Water is the most essential as well as the most abundant ingredient of life. Just as 70% of the earth's surface is covered with water, 70% of the human body is constituted of water. Eventhough the value and sanctity of water has been well known for thousands of years in eastern cultures, scientists in the west are only now beginning to break out of the "Chemicals are Chemicals" mode and examine the concept that water has memory, and that numerous intangibles (most notably the pathway and intention behind human intervention) are important factors in defining the value of water (Islam, 2006b).

At the industrial/commercial level however, preposterous treatment practices such as the addition of chlorine to 'purify'; the use toxic chemicals (soap) to get rid of dirt (the most potent natural cleaning agent (Islam,2006b); the use of glycol (very toxic) for freezing or drying (getting rid of water) a product; use of chemical CO_2 to render water into a dehydrating agent (opposite to what is promoted as 'refreshing'), then again demineralization followed by the addition of extra oxygen and ozone to 'vitalize'; the list seems to continue forever. Similar to what happens to food products (we call that the degradation of the following Chemical Technology Chain: Honey → Sugar → Saccharine → Aspartame), the chemical treatment technique promoted as water purification has taken a spiral-down turn (Islam, 2005). Chlorine treatment of water is common in the west and is synonymous with civilization. Similarly, transportation in copper pipe and distribution through stainless steel (enforced with heavy metal), storage in synthetic plastic containers and metal tanks, and mixing of ground water with surface water (itself collected from 'purified' sewage water) are common practices in 'developed' countries. More recent 'innovations' such as Ozone, UV and even H_2O_2 are proving to be worse than any other technology. Overall, water remains the most abundant resource, yet 'water war' is considered to be the most certain destiny of the 21st century. What Robert Curl (a Novel Laureate in Chemistry) termed as a 'technological disaster', modern technology development schemes seem to have targeted the most abundant resource (Islam, 2006b).

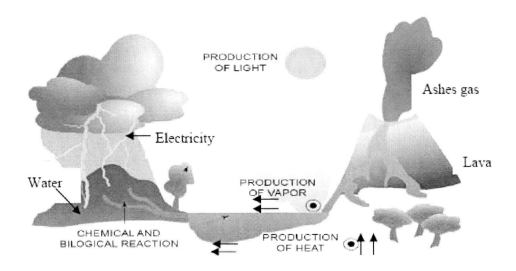

Figure 6. Natural transport phenomenon (after Fuchs, 1999)

Water vapor is considered to be one of the major greenhouse gases in the atmosphere. The greenhouse gas effect is thought to be one of the major mechanisms by which the radiative factors of the atmosphere influence the global climate. Moreover, the radiative regime of the radiative characteristics of the atmosphere is largely determined by some optically active component such as CO_2 and other gases, water vapor and aerosols (Kondratyev and Cracknell, 1998). As most of the incoming solar radiation passes through atmosphere and is absorbed by the Earth's surface, the direct heating of the surface water and evaporation of moisture results in heat transfer from the Earth's surface to the atmosphere. The transport of heat by the atmosphere leads to the transient weather system. The latent heat released due to the condensation of water vapors and the clouds play an important role in reflecting incoming short-wave solar radiation and absorbing and emitting long wave radiation. Aerosols such as volcanic dust and the particulates of fossil fuel combustion are important factors in determining the behavior of the climate system. Kondratyev and Cracknell (1998) reported that the conventional theory of calculating global warming potential only account for CO_2 ignoring the contribution of water vapor and other gases in global warming. Their calculation scheme took into account the other components affecting the absorption of radiation including CO_2, water vapor, N_2, O_2, CH_4, NOx, CO, SO_2, nitric acid, ethylene, acetylene, ethane, formaldehyde, chlorofluorocarbons, ammonia and aerosol formation of different chemical composition and various sizes. However, this theory fails to explain the effect of pure water vapour and the water vapour that is contaminated with chemical contaminants.

The impact of water vapour on climate change depends on the quality of water evaporated, its interaction with the atmospheric particulates of different chemical composition and size of the aerosols. There are at least 85,000 synthetic chemicals being used regularly throughout the world (Global pesticide campaigner, 2004; Icenhower, 2006). It has further been estimated that more than 2000 chemicals are introduced every year. Billions of tons of fossil fuels are consumed each year to produce these chemicals which are the major sources of water and air contamination. The majority of these chemicals are very toxic and radioactive, the particulates being continuously released into the atmosphere. The chemicals also reach to water bodies by leakage, transportation loss and as by-products of pesticides, herbicide and water disinfectants. The industrial wastes, which are contaminated with these chemicals finally reach to the water body and contaminate the entire water system. The particulates of these chemicals and aerosols when mixed with water vapour may increase the absorption characteristics in the atmosphere thereby increasing the possibility of trapping more heat. However, pure water vapour is one of the most essential components of the natural climate system and will have no impacts in global warming. Moreover, most of the water vapours will end up transforming into rain near the Earth's surface and will have no effect on the absorption and reflection. The water vapour in the warmer part of the earth could rise to higher altitudes as they are more buoyant. As the temperature decreases in the higher altitude, the water vapour gets colder, and it will hold less water vapor, reducing the possibility of increasing global warming.

Water is considered to have memory (Sudan, 1997; Tschulakow et al., 2005). Because of this property, the assumption of the impact of water vapour to global warming cannot be explained without the knowledge of memory. The impact will depend on the pathway it travels before and after the formation of vapour from water. Gilbert and Zhang (2003) reported that the nanoparticles change the crystal structure when they are wet. The change of structure taking place in the nanoparticles in the water vapour and aerosols in the atmosphere have profound impact in the climate change. This relation has been explained based on the memory characteristics of water and its pathway analysis. It is reported that water crystals are entirely sensitive to the external environment and take different shape based on the input (Emoto, 2004). Moreover, the history of water memory can be traced by its pathway analysis. The memory of water might have a significant role to play in the technological development (Hossain and Islam, 2006). Recent attempts have been directed towards understanding the role of history on the fundamental properties of water. These models take into account the intangible properties of water. This line of investigation can address the global warming phenomenon. The memory of water not only has impacts on energy and ecosystems but also has a key role to play in the global climate scenario.

CHARACTERIZATION OF ENERGY SOURCES

Various energy sources are classified based on a set of newly developed criteria. Energy is conventionally classified, valued or measured based on the absolute output from a system. The absolute value represents the steady state of energy source. However, modern science recognizes that such a state does not exist and every form of energy is at a state of flux. Various energy sources are characterized based on their pathways. Each form of energy has a set of characteristics features. Any time these features are violated through human intervention, the quality of the energy form declines. This analysis enables one to assign greater quality index to a form of energy that is closest to its natural state. Consequently, the heat coming from wood burning and the heat coming from electrical power will have different impacts on the quality, the only thing that comes in contact. Just as all chemicals are not the same, different forms of heat coming from different energy sources are not the same. The energy sources are classified based on the global efficiency of each technology, the environmental impact of the technology and overall value of energy systems (Chhetri et al., 2006c). The energy sources are classified based on the age of the fuel source in nature as it is transformed from one form to another (Chhetri et al., 2006a).

Various energy sources are also classified according to their global efficiency. Conventionally, energy efficiency is defined for a component or service as the amount of energy required in the production of that component or service; for example, the amount of cement that can be produced with one billion Btu of energy. Energy efficiency is improved when a given level of service is provided with reduced amounts of energy inputs, or services or products are increased for a given amount of energy input. However, the global efficiency of a system is defined as the efficiency calculated based on the energy input, products output, the possibility of multiple use of energy in the system, the use of the system by products and its impacts to the environment. The global efficiency calculation considers the source of the fuel, the pathways the energy system travels, conversion systems and impacts to the human health and environment and intermediate as well as by products of the energy system. Farzana and Islam (2006) calculated the global efficiency of various energy systems (Figure 7). They showed that global efficiencies of higher quality energy sources are higher than those of lower quality energy sources. With their ranking, solar energy source (when applied directly) is the most efficient, while nuclear energy is least efficient among many forms of energy studied. They demonstrated that previous finding fail to discover this logical ranking because the focus had been on local efficiency. For instance, nuclear energy is generally considered to be highly efficient which is a true observation if one's analysis is limited to one component of the overall process. If global efficiency is considered, of course the fuel enrichment alone involves numerous centrifugation stages. This enrichment alone will render the global efficiency very low. As an example, the global efficiency of a wood combustion process is presented.

Combustion of wood in traditional stoves has relatively low efficiency in the ranges of 14 % (Shastri et al., 2002). Chhetri (1997) reported from the experimental investigation that some of the stoves reached the efficiency of up to 20%. Some improved cookstoves have efficiency up to 25% (Kaoma and Kasali, 1994). However, the conventional efficiency calculation is based on calculating the local efficiency considering only the energy input and heat output in the system. This method does not consider the utilization of by-products, the fresh CO_2 which is essential for the plant photosynthesis, use of exhaust heat for household water heating, use of ash as surfactant in the enhanced oil recovery, fertilizer and good sources of natural minerals such silica, potassium, sodium, calcium and others. This wood ash can be used as catalysts in different chemical processes such as replacement of synthetic potassium and sodium hydroxides (Rahman et al., 2004; Rahman and Islam, 2006).

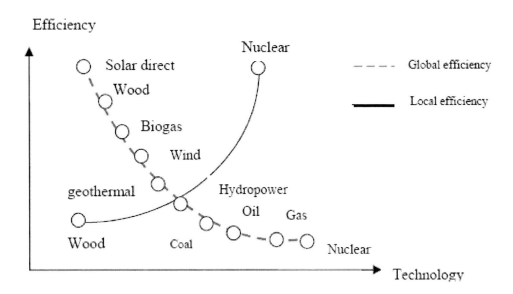

Figure 7. Global and local efficiency of different energy sources

Wood ash is very rich source of silica which is an important source for industrial applications. The ash also contains various minerals such as potassium, sodium, magnesium, calcium and others. Conventionally, ash has been in use as a source of fertilizer because of its high mineral content. A good wood ash is also a truly natural detergent. Sodium or potassium can also be extracted from wood ash to use as saponification agent for soap making from vegetable oils and animal or fish fats. A fine wood ash is an effective raw material for manufacturing numerous health products. Chhetri and Islam (2006a) reported that wood ash can be extracted to use as natural catalyst for the transesterification of vegetable oil to produce biodiesel for diesel substitute.

Rahman et al. (2004) reported that Maple wood ash has the potential to adsorb both Arsenic (III) and Arsenic (V) from contaminated aqueous streams at low concentration levels without any chemical treatment. Static tests showed an arsenic removal up to 80% and in various dynamic column tests, the arsenic concentration was reduced from 500 ppb to lower than 5ppb. Moreover, ash had been in use traditionally as a water disinfecting agent possibly because of some mineral content in it. Khan et al. (2006c) developed an energy efficient stove fuelled by compacted saw dust that utilizes exhaust heat coming from flue gas for household water heating. A simple heat exchanger can be used to transfer heat from flue gas to cold water. They also designed an oil-water trap to trap all particulate matters emitted from wood combustion. The particulates or the carbon soot collected from oil-water mixture is a very good source of nano-materials which have a very high industrial demand. The carbon soot can also be used as a non toxic paint.

When the particulates are trapped into oil-water trap and the heat is extracted for water heating, the CO_2 emitted is a 'fresh' or 'new' or 'natural' CO_2 which is most favored by plants during photosynthesis (Islam, 2005; Chhetri et al, 2006a). Combustion of woodfuel thus does not contribute to the greenhouse effect. Other products of combustion such as NOx and formaldehyde are also not harmful compared to the similar products from fossil fuel burning. The heat loss in the heater itself will contribute to some loss of efficiency. Assuming the 5-10% radiation and conduction loss, the efficiency of wood combustion in the stove is considered to be approximately 90%. Thus, wood combustion in effectively designed stoves has one of the highest efficiency among the other technology types. Figure 6 shows the classification of energy sources based on their global efficiency. Based on the global efficiency, the nuclear energy has the lowest efficiency. Direct solar application has the highest efficiency among the energy sources because the solar energy source is free and has no negative environmental impacts.

THE KYOTO PROTOCOL

Various climate models using different scenarios have been analyzed. Emission scenarios under satisfactory as well as partial fulfillment of Kyoto Protocol have been evaluated. Based on the conclusion that current global warming and climate change is due to the emission of greenhouse gases by industrial activities, the Kyoto Protocol was negotiated by more than 160 nations in 1997, aiming to reduce greenhouse gases primarily CO_2 (EIA, 2006). In this Protocol, the industrial nations (Annex-I countries) have committed to making substantial reductions in their emission of greenhouse gases by 2012. For the first time, the Kyoto Protocol includes an international agreement for the reduction of emission of greenhouse gases. Global warming is a major environmental concern. It is especially the case in many developed countries, where the greenhouse gas emissions responsible for this change are concentrated. As a result, uncertainties and fears rise about possible consequences for the development of manufacturing activities in the future. According to the Third Assessment Report of the Intergovernmental Panel on Climate Change (IPPC) which brings together the world's leading experts in this field, globally averaged surface temperature is projected to increase by between minimum of 1.4^0C and maximum of 5.8°C from 1990 to 2100 under a business-as-usual projection. This temperature rise corresponds to a sea level rise of 9 cm to a maximum of 88 cm. More recently, chief scientific advisor to UK government estimated an increase of 3^0C in the coming decade. Despite desserting views and skepticism (Lindzen, 2002; Lindzen, 2006;Carter, 2006) it is being increasingly clear that global warming is not a natural phenomenon and emerges from industrial practices that are not sustainable (Khan, 2006).

The Kyoto Protocol has set special targets in order to reduce greenhouse gas emission from Annex-I countries as outlined by Article 2 of the Protocol (Kyoto Protocol, 1997). Various measures suggested reducing the greenhouse gas emissions are promotion of sustainable development, enhancement of energy efficiency, protection and enhancement of sinks and reservoirs of greenhouse, promotion of sustainable forms of agriculture in light of climate change considerations, methane reduction through waste management and promotion of increased use of new and renewable forms of energy. Article 3 as major provision of the Kyoto Protocol reflects agreement that all parties must reduce their greenhouse gas emission level 5% below their 1990 levels during the 'commitment period' between 2008 and 2012. Some countries, including the US, Canada, European Union countries and Japan, will have to reduce emissions up to 8 percent below 1990 level. Some on this list including Australia and Iceland will be allowed to increase emissions by varying amounts up to 10%. The other features of the Protocol is that the parties included in Annex B may participate in emissions trading for the purposes of fulfilling their commitments. Clean Development Mechanism

(CDM) is defined under article 12 to assist sustainable development for developing countries. Annex I countries could count reductions in greenhouse gases achieved in this way against their own targets.

Despite a series of targets for emission reduction, the Kyoto Protocol has many flaws. The standards for emission reduction are not based on scientific facts. The time scale proposed for reducing emission levels also had no justification. The standards emission level taken as of 1990 had also no basis. Developing countries such as China and India as newly emerging economies were excluded from meeting the targets. Emission trading has become 'license to pollute' for industrialized countries and big corporations. This situation has become worse due to the introduction of emission trading. The CDM instituted to assist developing countries is not functional. Moreover, there are significant bureaucratic formalities that slow down the project approval for CDM granting to develop clean energy technologies. The main difficulties in making the CDM work arise from the dual issues of 'additionality' and 'baselining'. To obtain a 'certified emission reduction', that is, new emission rights, from investments in developing countries, investors must demonstrate that emission reductions are 'additional' to any that would occur in the absence of the certified project activities. As a result, only 49 projects out of 1030 submitted were approved by 2002 (Pershing and Cedric, 2002). The monitoring and administration of such emission certification would be a cost burden for relatively small companies. As a result, only the big companies which can administer and monitor the certification would get the benefits. The present targets of greenhouse gas emission in industrialized countries are not tough enough for 2008-2012. Such provisions are affected by many factors and would likely become a license to pollute that will enable global emission to further increase. The Kyoto Protocol doesn't even set a long-term goal for atmospheric concentrations of CO_2, so the Kyoto protocol does not hold good promise to achieve its set targets. Possibly the most important shortcoming of the Kyoto protocol is in its failure to recommend any change in the current process of energy production and utilization. Any change in the current practice can alter the global warming scenario drastically. For instance, if toxic chemicals were not used in crude oil refining, allowable CO_2 emission would increase very significantly. Such analysis is absent in most of the previous work (Khan 2006; Khan and Islam, 2006b).

The intergovernmental Panel on Climate Change stated that there was a "discernible" human influence on climate; and that the observed warming trend is "unlikely to be entirely natural in origin" (IPCC, 2001a). The Third Assessment Report of IPCC stated "There is new and stronger evidence that most of the warming observed over the last 50 years is attributable to human activities." Khilyuk and Chilingar (2004) reported that the CO2 concentration in the atmosphere between 1958 to 1978 was proportional to the CO2 emission due to the burning of fossil fuel. In 1978, CO2 emissions into the atmosphere due to fossil fuel burning stopped rising and were stable for nine years. They concluded that if fossil fuels burning were the main cause, the atmospheric concentration should stop rising and thus, fossil fuel burning is not the cause of the greenhouse effect. However, this assumption is extremely short sighted and global climate certainly does not work linearly, as envisioned by Khilyuk and Chilingar (2004). Moreover, the 'Greenhouse Effect One-Layer Model' proposed by Khilyuk and Chilingar (2003; 2004) assumes adiabatic conditions in the atmosphere that do not practically exist. The authors have concluded that the human-induced emission of carbon dioxide and other greenhouse gases have a very small effect on global warming. This is due to the limitation of the current linear computer models which cannot predict temperature effects on the atmosphere other than the low level. Similar arguments were made while promoting dichlorodifluoromethane (CFC-12) to environmental problems incurred by ammonia and other refrigerants after decades of use, CFC-12 was banned in USA in 1996 for its impacts on stratospheric ozone layer depletion and global warming. Khan and Islam (2006b) presented detailed list of technologies that were based on spurious promises. Zatzman and Islam (2006) complemented this list by providing a detailed list of economic models that are also counter productive. The potential impact of microbial activities on the mass and content of gaseous mixtures in Earth's atmosphere on a global scale was explained (Khilyuk and Chilingar, 2004). However, this study does not distinguish between biological sources of greenhouse gas emission such as from microbial activities, and industrial sources such as fossil fuel burning. They inhibit different characteristics as they derive from diverse origins and travel different paths which obviously, have significant impact on atmospheric processes.

Current climate models have several problems. Scientists have agreed on the likely rise in the global temperature over the next century. However, the current global climatic models can predict only global average temperatures. Projection of climate change in a particular region is considered to be beyond current human ability. Atmospheric Ocean General Circulation Models (AOGCM) are used by IPCC to model climatic feactures, however, these models are not accurate enough to provide reliable forecast on how climate may change. They are linear models and cannot forecast complex climatic features. Some climate models are based on CO_2 doubling and transient scenarios. However, the effect of climate while doubling the concentration of CO_2 in the atmosphere cannot predict the climate in other scenarios. These models are insensitive to the difference between natural and industrial greenhouse gases. There are some simple models in use which use fewer dimensions than complex models and do not predict complex systems. The Earth System Models of Intermediate Complexity (EMIC) are used to bridge the gap between the complex and simple models, however, these models are not suitable to assess the regional aspect of climate change (IPCC, 2001a; IPCC, 2001b).

Unsustainable technologies are the major cause of global climate change. Sustainable technologies can be developed following the principles of nature. In nature, all functions are inherently sustainable, efficient and functional for an unlimited time period. In other words, as far as natural processes are concerned, 'time tends to Infinity'. This can be expressed as t or, for that matter, $\Delta t \rightarrow \infty$. By following the same path as the functions inherent in nature, an inherently sustainable technology can be developed (Khan and Islam, 2006b). The 'time criterion' is a defining factor in the sustainability and virtually infinite durability of natural functions. Figure 8 shows the direction of nature-based, inherently sustainable technology, as contrasted with an unsustainable technology. The path of sustainable technology is its long-term durability and environmentally wholesome impact, while unsustainable technology is marked by Δt approaching 0. Presently, the most commonly used theme in technology development is to select technologies that are good for t='right now', or $\Delta t=0$. In reality, such models are devoid of any real basis termed "aphenomenal" by Khan et al., (2005c) and should not be applied in technology development if we seek sustainability for economic, social and environmental purposes. While developing the technology for any particular climatic model, this sustainability criterion is truly instrumental. The great flaw of conventional climate models is that they are focused on the extremely short term, t='right now', or $\Delta t=0$.

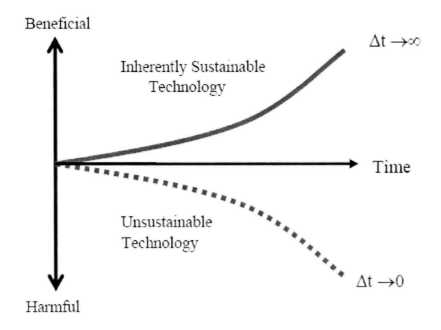

Figure 8. Direction of sustainable/green technology (redrawn from Islam, 2005)

SUSTAINABILE ENERGY DEVELOPMENT

Different technologies that are sustainable for long term and do not produce any greenhouse gases are presented. Technology has a vital role to play in modern society. One of the major causes of present day environmental problems is the use of unsustainable technologies. The use of thousands of toxic chemicals in fossil fuel refining, industrial processes to the products of personal care such as body lotion, cosmetics, soaps and others has polluted much of the world in which we live (The Globe and Mail 2006; Chhetri et al., 2006b). Present day technologies are based on the use of fossil fuel in the form of primary energy supply, production or processing, and the feedstock for products, such as plastic. Every stage of this development involves the generation of toxic waste, rendering product harmful to the environment. According to the criterion presented by Khan and Islam (2005 a; 2005 b), toxicity of products mainly comes from the addition of chemical compounds that are toxic. This leads to continuously degrading quality of the feedstock. Today, it is becoming increasingly clear that the "chemical addition" that once was synonymous with modern civilization is the principal cause of numerous health problems including cancer and diabetes. A detailed list of these chemicals has been presented by Khan and Islam (2006 b).

Proposing wrong solution for various problems has become progressively worse. For instance, USA is the biggest consumer of milk, most of which is 'fortified' with calcium. Yet USA ranks at the top of the list of osteoporosis patients per capita in the world. Similar standards are made about the use of vitamins, antioxidants, sugarfree diet etc. Potato farms on Prince Edward Island in eastern Canada are considered a hot bed for cancer (The Epoch Times, 2006). Chlorothalonil, a fungicide, which is widely used in the potato fields, is considered a carcinogen. US EPA has classified chlorothalonil as a known carcinogen that can cause a variety of ill effects including skin and eye irritation, reproductive disorders kidney damage and cancer. Environment Canada (2006) published lists of chemicals which were banned at different times. This indicates that all the toxic chemicals used today are not beneficial and will be banned from use some day. This trend continues for each and every technological development. However, few studies have integrated these findings to develop a comprehensive cause and effect model. This comprehensive scientific model developed by Khan and Islam (2006b) is applied for screening unsustainable and harmful technologies right at the onset. Some recently developed technologies that are sustainable for the long term are presented.

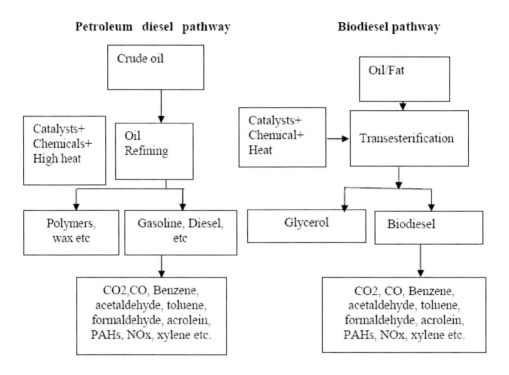

Figure 9. Pathway of mineral diesel and conventional biodiesel (Chhetri and Islam, 2006a)

One of the sustainable technologies presented in this paper is the true green biodiesel model (Chhetri and Islam, 2006a). As an alternative to petrodiesel, biodiesel is a renewable fuel that is derived from vegetable oils and animal fats. However, the existing biodiesel production process is neither completely 'green' nor renewable because it utilizes fossil fuels, mainly natural gas, as an input for methanol production. Conventional biodiesel production process entails the use of fossil fuels such as methane as an input to methanol. It has been reported that up to 35% of the total primary energy requirement for biodiesel production comes from fossil fuel (Carraretto et al., 2004). Methanol makes up about 10% of the feed stock input and since most methanols are currently produced from natural gas, biodiesel is not completely renewable (Gerpen et al., 2004). The catalysts and chemicals currently in use for biodiesel production are highly caustic and toxic. The synthetic catalysts used for the transesterification process are sulfuric acid; sodium hydroxide and potassium hydroxide, which are highly toxic and corrosive chemicals. The pathway for conventional biodiesel production and petrodiesel production follows a similar path (Figure 9). Both the fuels have similar emission of pollutants such as benzene, acetaldehyde, toluene, formaldehyde, acrolein, PAHs, xylene (EPA, 2002). However, the biodiesel has fewer pollutants in quantity than petrodiesel.

Chhetri and Islam (2006a) developed a process that rendered the biodiesel production process truly green. This process used waste vegetable oil as biodiesel feedstock. The catalysts and chemicals used in the process were non-toxic, inexpensive and natural catalysts. The catalysts used were sodium hydroxide obtained from the electrolysis of natural sea salt and potassium hydroxide from wood ash. The new process substituted the fossil fuel based methanol with ethanol produced by grain based renewable products. Use of natural catalysts and non toxic chemicals overcame the limitation of the existing process. Fossil fuel was replaced by direct solar energy for heating, making the biodiesel production process independent of fossil fuel consumption.

Khan et al. (2006c) developed a criterion to test the sustainability of the green biodiesel. According to this criterion, to consider any technology sustainable in the long term, it should be environmentally appealing, economically attractive and socially responsible. The technology should continue for infinite time maintaining the indicators functional for all time horizons. For a green biodiesel, the total environmental benefits, social benefits and economics benefits are higher than the input for all time horizons. For example, in case of environmental benefits, green biodiesel burning produces 'natural' CO_2 which can be readily tahen up by plants. The formaldehyde produced during biodiesel burning is also not harmful as there are no toxic additives involved in the biodiesel production process. The plants and vegetable for biodiesel feedstock production also have positive environmental impacts. Thus switching from petrodiesel to biodiesel fulfils the condition $\frac{dCn_t}{dt} \geq 0$ where Cn is the total environmental capital of life cycle process of biodiesel production. Similarly, the total social benefit (Cs) $\frac{dCs_t}{dt} \geq 0$ and economic benefit (Ce) $\frac{dCe_t}{dt} \geq 0$ by switching from mineral diesel to biodiesel (Khan et al., 2006b). Figure 10 gives a sustainable regime for an energy system for infinite time and fulfills the environmental, social and economic indicators. Biodiesel can be used in practically all areas where petrodiesel is being used. This substitution will help to significantly reduce the CO_2 responsible for current global warming problem.

Bioethanol is another sustainable technology that offers replacement for gasoline engines. The global gasoline consumption is approximately 12 billion liters per year (Martinot, 2005). This is one of the major sources for CO_2 emission. Current gasoline replacement by bioethanol fuel is approximately 32 billion liters worldwide. The conventional bioethanol production from various feedstocks such as switchgrass and other biomass involves use of chemicals for its breakdown in various stages. For example, the ethanol production process from switchgrass involves acid hydrolysis as a major production process. It is reported that the conversion of switchgrass into bioethanol use concentrated sulfuric acid at 4:1 (acid biomass ratio) which makes the process unsustainable; produced fuel is a highly toxic fuel and produces formentation inhibitors such as 5-hydroxymethylfurfural (5-HMF) and furfural acid during the hydrolysis process which reduces the efficiency (Bakker et al, 2004). Moreover, the conventional bioethanol production also consumes large amount of fossil fuel as primary energy input making the ethanol production dependent on fossil fuels.

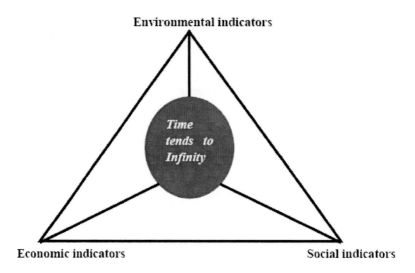

Figure 10. Major elements of sustainability in technology development (After Khan at el., 2006b)

Development of bioenergy on a large scale requires the deployment of environmentally acceptable, low cost energy crops as well as sustainable technologies to harness them with the least environmental impact. Sugarcane, corn, switchgrass and other ligocellulogic biomass are the major feed stocks for ethanol production. Chhetri (2006) developed a process that makes the bioethanol production process, truly green. They proposed the use of non toxic chemicals and natural catalysts to make the bioethanol process truly environmental friendly. The technology has been tested for long term sustainability using a set of sustainability criteria. The ethanol produced using the natural and non toxic catalysts will produce natural CO_2 after combustion and has no impacts on global warming.

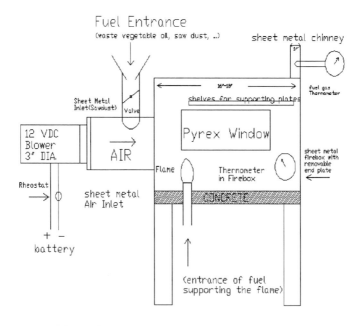

Figure 11. Saw dust to electricity model jet engine (Vafaei, 2006)

Recently, a jet engine has been designed in order to convert saw dust waste to electricity (Vafaei, 2006). This is one of the most efficient technologies since it can use a variety of fuels for combustion. This was designed primarily to use saw dust to produce power for the engine (Figure 11). In this jet engine, saw dust is sprayed from the top where the air blower works to make a jet. Some startup fuel such as organic alcohol is used to start up the engine. Once the engine is started, the saw dust and blower will be enough to create power for the engine to run. The main advantage of this jet engine is that it can use a variety of fuels such as waste vegetable oil and tree leaves. It has been reported that crude oil can be directly burnt in such engines (Vafaei, 2006). The utilization of waste saw dust and waste vegetable oil increases the global efficiency of the systems significantly. The possibility of directly using of crude oil can eliminate the various toxic and expensive refining processes which alone reduce large amounts of greenhouse gas emission into the atmosphere. This technology is envisaged as one of the most sustainable technologies among the others currently available.

ZERO WASTE ENERGY SYSTEMS

Different zero waste technologies that eliminate the production of industrial CO_2 are described. Modern civilization is synonymous with the waste generation (Islam, 2003; Islam, 2004; Khan and Islam, 2006b). This trend has the most profound impact on energy and mass utilization. Conventional energy systems are most inefficient technologies (Farzana and Islam, 2006). The more is wastes, the more inefficient the system is. Almost all industrial and chemical processes produce wastes and most of them are toxic. The wastes not only reduce the efficiency of a system but also pose severe impacts on health and environment leading to further degradation of global efficiency. The treatment of this toxic waste is also highly expensive. Plastics derivatives from refined oil are more toxic than original feedstocks; the oxidation of a plastic tire at high temperature produces toxics such as dioxin. The more refined the products are, the more wastes are generated. A series of zero waste technologies are presented. They are analogous to the 'five zeros' of Olympic logo which are zero emissions, zero resource waste, zero waste in activities, zero use of toxics and zero waste in the product life cycle. This model, originally developed by Lakhal and H'Midi (2003) was called the Olympic Green Chain model. This model was used by Khan (Ph.D., thesis) for proposing an array of zero waste technologies.

The utilization of waste as an energy source offers multiple solutions for energy as well as mitigation of environmental problems. Production of energy from waste reduces the cost of waste treatment and at the same time gives added value to the waste products. This enhances the global efficiency of the system. Local efficiency is calculated based on the output and input of the turbine and generating engine. This method does not consider the exploration and processing, total transmission and distribution losses, the environmental and social cost associated with the system and the possible uses of by-products. In the case of global efficiency, the efficiency of fuel exploration and processing, efficiency of all moving parts such as turbine and generator, transmission and distribution losses, the embodied energy cost and emission during parts manufacturing, the total impact on environment and health, including biodiversity due to the technological intervention, are considered.

The global efficiency of coal burning to produce electricity is calculated considering the efficiency of the entire operations from coal mining to electricity transmission. The coal to electricity production without considering the environmental impacts, has a global efficiency of 12.40% (Farzana and Islam, 2006). Figure 12 is the schematic of calculating the global efficiency of an energy system. The efficiency can be improved by using the fly ash for other purposes such as making cement, if toxic catalysts are not used during coal conversion or cracking. Some chemicals such as of Fe-Mn oxides or some acidic compounds are added during coal cleaning and refining before they are burned (Guo et al., 2004). These additives contaminate the CO_2 making it toxic. A large amount of sulfur and arsenic are also released during coal burning. The global efficiency of the whole system is thus reduced even if we consider the environmental impacts of coal burning.

The global efficiency of wood combustion has been calculated to be more than 90% (Chhetri et al., 2006a). Similarly, the global efficiency of nuclear power generation was approximately 5%. Nuclear waste storage has been the subject of discussion for a long time and no feasible methods have been successfully worked out yet. Development of nuclear power generation also utilizes fossil fuel during various process operations contributing to the emission of CO_2. The major problem of nuclear power generation is that the half lives of natural uranium isotopes U-234 is 244,500 years, U-235 is 7.03×10^8 years and that of U-238 is 4.46×10^9 years (Wise Uranium Project, 2005). The half lives of the 'enriched' uranium is much more higher than this. Because of these reasons, nuclear power has the least global efficiency. Nuclear waste is a big problem which cannot be utilized and is almost impossible to store safely.

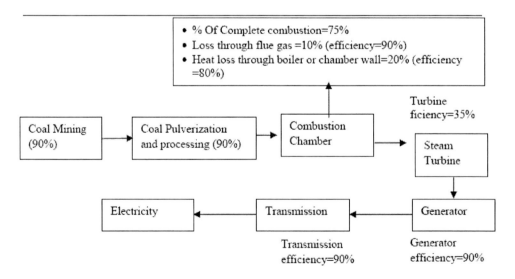

Figure 12. Global efficiency of coal to electricity system (Farzana and Islam, 2006)

Solar energy is free energy, extremely efficient. Direct solar energy is a benign technology. Khan et al. (2005b) developed a direct solar heating unit to heat the waste vegetable oil as heat transfer medium (Figure 13). The solar concentrator can heat the oil to more than 300^0C and the heat can be transferred through heat exchanger for space heating, water heating or any other purposes. In conventional water heating, the maximum heat can be stored is 100^0C. However, in the case of direct oil heating, the global efficiency of the system is more than 80%. No waste is generated in the system.

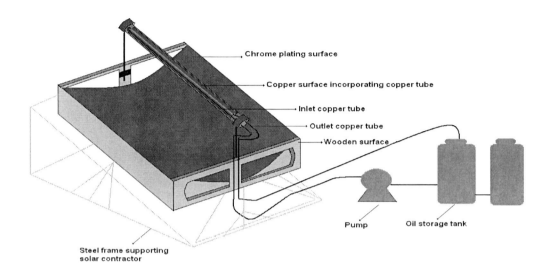

Figure 13. Details of solar heating unit (After Khan et al., 2006b)

Khan et al. (2006a) developed a heating/cooling and refrigeration system that uses direct solar heat without converting into electricity. The single pressure refrigeration cycle is a thermally driven cycle that uses three fluids. One fluid acts as a refrigerant, the second as a pressure-equalizing fluid, and a third as an absorbing fluid. Because the cycle operates at a single pressure, no moving parts, such as a pump or compressor, are required. In order to facilitate fluid motion, the cycle uses a bubble pump that uses heat to ensure the drive. All of the energy input is in the form of heat. Utilization of direct heat for heating cooling or refrigeration replaces the use of large amounts of fossil fuel reducing the CO_2 emission significantly. This type of refrigerator has silent operations, higher heat efficiency, no moving parts and portability.

Khan et al., (2006c) developed a novel zero waste saw dust stove in order to utilize the waste saw dust (Figure 14). At present saw dust is considered a waste, and management of waste always involves cost. Utilizing the waste saw dust enhances the value addition of the waste material and generates valuable energy as well.

The particulates of the wood burning are collected in an oil-water trap which is a valuable nano-material for many industrial applications. This soot material can also be used as an ingredient for non toxic paint. Figure 15 is the schematic of zero waste saw dust stove. The fuel is waste material from industries as an input. The stove generates energy mainly for cooking. The flue gas heat can be extracted for water heating and the CO_2 from the saw dust burning is a natural CO_2 for plant photosynthesis. The ash collected from the ash port of the stove can be used as fertilizer and cleaning agents as well as replacement of synthetic sodium or potassium hydroxide for industrial application. This is totally zero waste technology. Since half of the world's population uses domestic cooking stoves for cooking and space heating, dissemination of such stoves would significantly contribute to replace LPG gas stoves and other petroleum based fuel.

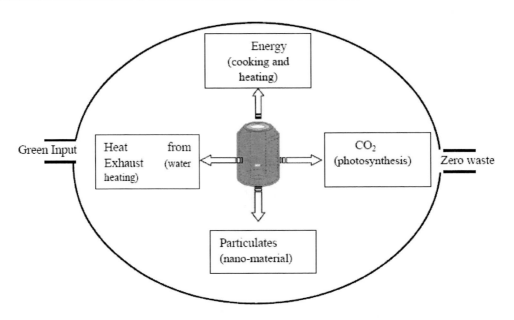

Figure 14. Pictorial view of zero-waste models for wood stove (after Khan et al., 2006c)

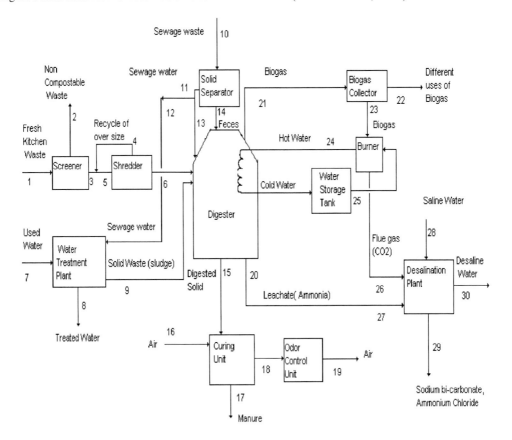

Figure 15. Zero-waste mass utilization scheme

A high efficiency jet engine has also been developed (Vafaei, 2006). This jet engine can use practically any type of solid fuel such as waste sawdust, and liquid fuel such as waste vegetable oil and crude oil to run the engine. A jet is created to increase the surface area in order to increase the burn rate. Direct combustion of crude oil is possible in this engine. This development will eliminate the costly refining and processing of petroleum products. This will have a significant impact in reducing industrial CO_2 as the refining process is what makes the CO_2 a toxic product, due to the use of heavy metals and toxic catalysts.

Khan et al. (2005b) proposed an approach for zero-waste (mass) utilization for typical urban setting including processing and regeneration of solid, liquid and gas. In this process, kitchen waste and sewage waste are utilized for various purposes including biogas production, desalination, water heating from flue gas and good fertilizer for agricultural production. The carbon dioxide generated from biogas burning is utilized for desalination plant. This process achieves zero-waste in mass utilization. The process is shown in Figure 15. The technology development in this line has no negative impact on global warming.

REVERSING GLOBAL WARMING-ROLE OF TECHNOLOGY DEVELOPMENT

A series of techniques are discussed to reduce industrial CO_2 which contributes to global warming. Conventional energy sources, such as fossil fuel contribute to greenhouse gases emissions. Because various toxic chemicals and catalysts are used for refining/processing oil and natural gas, the emitted CO_2 is a toxic product. In addition, fossil fuels have greater properties of carbon isotope ^{13}C making them mere likely to be readily absorbed by plants. This leads to the alteration of characteristic recycle period of carbon dioxide, causing delays that result in an increase in total CO_2 in the atmosphere. Billions of people in the world use traditional stoves fueled by biomass, for their cooking and space heating requirement. It is widely held that wood burning stoves emit more pollution into the atmosphere compared with oil and natural gas burning stoves. However, a small intervention in wood burning stoves will result in the emission of natural CO_2, which is essential for natural processes. Identifying the limitations of conventional stoves, a new technique has been developed to achieve zero-waste in such technologies. This line of development will have great impact on technological development in industrial and other sectors as well.

Figure 16 is a saw dust packed cookstove developed by Khan et al., (2006c). This is a highly efficient clay stove that has no waste. Even though not all by-products are captured and value added, this stove is still considered to be a zero waste stove as it produces only organic gases that are readily absorbed by the environment in order to produce useful final products. This technology offers solutions for the production of a natural form of CO_2 which is readily synthesized by plants. The other emissions such as methane and oxides of nitrogen are not harmful, unlike those emitted from petroleum based fuels.

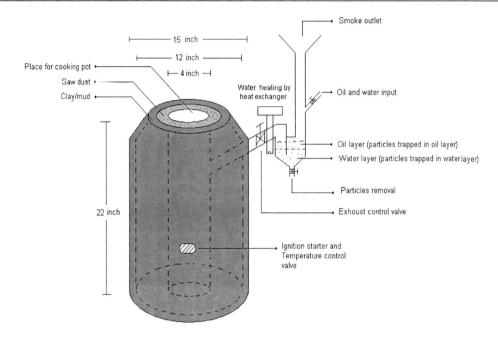

Figure 16. Sawdust packed stove (after Khan et al, 2006c)

The production of green biodiesel and bioethanol discussed earlier are key elements in the production of natural CO_2, which does not contribute to the global warming. Only non-toxic chemicals and catalysts are used in the processes to produce biodiesel and bioethanol. Even the benzene, NO_x, methane and other emissions are not harmful as there is no toxic chemicals are involved during the production. These fuels are derived from renewable sources such as plants and vegetables. Plant and vegetables are essential components of natural food cycles for both the plants and animal kingdoms. It has been reported that petroleum fuels are being exhausted within the next few decades, whereas renewable biofuel sources continue for infinite time. These biofuels could replace all petroleum fuels provided the biomass farming is planned in a sustainable way. Replacing the petroleum fuels with clean biofuels in a sustainable way, can eventually lead to the reversal of global warming. This reversal can be accelerated if processing of fossil fuel is rendered non toxic by avoiding the use of toxic additives (Khan and Islam, 2006b; Al Darby et al., 2002). Recent studies indicated that crude oil refining can be avoided altogether with some design modification of the combustion engine (Vafaei, 2006)

Energy systems are classified based on their global efficiencies. According to Farzana and Islam (2006), global efficiency is one of the major indicators to be considered while selecting the technology choice for any energy systems. For instance, conventional oil heated steam turbine to electricity production, has a global efficiency of approximately 16%. For combined heat and power turbines, the global efficiency is approximately 18%. Similarly, the global efficiency of a coal fired power system is approximately 15%, for hydropower systems it is 43%, from biomass to electricity conversion is 13%, and nuclear power plant has a global efficiency of approximately 5%. The environmental cost due to these technologies has not been added up yet which further reduces the global efficiency of these technologies. The solar photovoltaic conversion efficiency is reported to be around 15% (Islam et al., 2006). This system also uses toxic batteries and synthetic silicon cells which includes toxic heavy metals inside solar cells. Their efficiency starts decreasing when we store them in the batteries. The efficiency of a battery itself is not very high and batteries contain toxic compounds in it. The inert gas filled tubes radiate very toxic light. The 15% global efficiency of a systems means that for every 100 units of energy produced, 85% energy is lost in the process. This indicates that the prevailing technologies have a very high waste generation. Most of these systems produce toxic CO_2 that cause the global warming problems. Moreover, the CO_2 emissions from embodied energy associated with all the massive equipment production and manufacturing is also very high.

Direct application of solar energy for heating, is of the highest efficiency. Solar energy is a free source and it has no negative environmental impact at all. Similarly, wood combustion in a simple stove has also a very high global efficiency due to the use of by-products and waste heat. The CO_2 from wood combustion is also an essential ingredient for maintaining the photosynthesis process. The energy systems which have highest efficiency have the lowest environmental impacts.

Considering the long term impacts of various energy systems, the CO_2 emissions during combustion of oil and gas, coal, embodied energy and associated CO_2 emission for hydropower and photovoltaic systems has been ranked based on the quality of CO_2. Similarly, the CO_2 emission from geothermal energy, biomass burning has also been ranked. Based on this ranking, natural CO_2 which does not contribute to global warming is deducted from the industrial CO_2 which contributed global warming. Series of technological interventions have been suggested that can eventually reverse the global warming problem.

CONCLUSION

It is concluded that the current synthetic chemical based technological development are major causes for global warming and climate change problems. Emission of industrial CO_2 that is contaminated by the addition of toxic chemicals during fuel refining, processing and production activities are responsible for global warming. The natural CO_2 which is not only beneficial to the environment but also an essential ingredient for life and biodiversity in the earth does not cause global warming. For the first time, natural and industrial CO_2 has been differentiated. Carbon dioxide is characterized based on various criteria such as the origin, the pathway it travels and isotope numbers. Current status of greenhouse gas emissions from various anthropogenic activities is discussed. Role of water in global warming has been detailed. Various energy sources are classified based on their global efficiencies. The assumptions and implementation mechanisms of the Kyoto Protocol have been critically reviewed and argued that Clean Development Mechanism of the Kyoto Protocol has become the 'license to pollute" due to its improper implementation mechanism. The conventional climatic models are deconstructed and guidelines for new models are developed in order to achieve the true sustainability in technology development in the long term. A series of sustainable technologies that produce natural CO_2 which do not contribute to global warming were presented. Various zero-waste technologies that have no negative impact on environment are keys to reverse the global warming. This paper shows that a complete reversal of current global warming problem is possible only if pro-nature technologies are developed.

REFERENCES

AlDarbi, M., Saeed, N.O., Islam, M.R. and Lee, K. 2005. Biodegradation of Natural Oils in Seawater. *Energy Sources.* 27(1,2):19-34.

AlDarbi, M., Muntasser, Z., Tango, M. and Islam, M.R., 2002. Control of Microbial Corrosion Using Coatings and Natural Additives. *Energy Sources* 24(11): 1009-1018

Agus, D. B. and Vera, J. C. and Golde, D. W. 1999. Stromal Cell Oxidation: A Mechanism by Which Tumors Obtain Vitamin C. *Cancer Research* 59:4555–4558

Bakker, R.R., Gosselink, R.J.A., Maas, R.H.W., Vrije, T. de and Jong, E. de. 2004. Biofuel Production from Acid-Impregnated Willow and Switchgrass. 2nd World Conference on Biomass for Energy, Industry and Climate Protection, 10-14 May 2004, Rome, Italy

Carter, B., 2006. There Is a Problem With Global Warming…It Stopped in 1998. The Telegraph,UK.www.telegraph.co.uk/opinion/main.jhtml?xml=/opinion/2006/04/09/do0907.xml (accessed on June 30, 2006)

Chhetri and Islam, 2006a. Towards Producing a Truly Green Biodiesel. Energy Sources: In Press.

Chhetri and Islam, 2006b. Role of CO_2 in Global warming: *Nat.Sci.and Sust.Dev. Paper in progress.*

Chhetri, A.B., 1997. An Experimental Study of Emission Factors from Domestic Biomass Cookstoves. AIT Master's Degree Thesis no- ET -97-34.

Chhetri, 2006. Scientific Characterization of Global Energy Sources. *J of Nat.Sci. and Sust.Tech.* Submitted.

Chhetri A.B., Khan, M.I. and Islam, M.R. 2006a. A Novel Sustainably Developed Cooking Stove. *Nat.Sci.and Sust.Dev.* Submitted.

Chhetri, A.B., Rahman, M.S. and Islam, M.R., 2006b. Production of Truly 'Healthy' Health Products. 2nd International Conference on Appropriate Technology, July 12- 14, 2006, Zimbabwe.

Chhetri, A.B., Zaman, M.S. and Islam, M.R., 2006c. Characterization of Energy Sources Based on Their Value. Paper in Progress.

Carraretto, C., Macor, A., Mirandola, A., Stoppato, A., Tonon S., 2004. Biodiesel as Alternative Fuel: Experimental Analysis and Energetic Evaluations. *Energy* 29: 2195–2211

Environment Canada, 2006. www.ec.gc.ca/international/multilat/rotterdam_e.htm

EIA, 2004. International Energy Outlook 2004", Greenhouse Gases general information, Energy Information Administration, Environmental Issues and World Energy Use. EI 30, 1000 Independence Avenue, SW, Washington, DC 20585.

EIA, 2006. International Petroleum Monthly, May 2, 2006.

Emoto, M., 2004. The messages from Water. Conscious Water Crystals: The Power of Prayer Made Visible. www.life-enthusiast.com/twilight/research_emoto.htm <accessed on June 1, 2006>.

EPA global warming site(2001)Climate.http://www.epa.gov/globalwarming/climate/index.htm

EPA, 2002. A Comprehensive Analysis of Biodiesel Impacts on Exhaust Emissions. *Air and Radiation.* Draft technical report. EPA420-P-02-001.

Fuchs, H.U. 1999. A Systems View of Natural Processes: Teaching Physics the System Dynamics Way. *The Creative Learning Exchange* 8 (1): 1-9

Farahani, S., Worrell, E. and Bryntse, G. 2004. CO2-free paper? Resources, Conservation and Recycling 42:317–336

Farquhar, G.D., Ehleringer, J.R. and Hubick, K.T. 1989. Carbon Isotope Discrimination and Photosynthesis. *Annu.Rev.Plant Physiol.Plant Mol. Biol.* 40:503-537

Farzana, N. and Islam, M.R., 2006. Global Efficiency of Energy Systems. *Energy sources.* Submitted.

Goldschmidt, V.W.2005. Two Differing Perspectives on Ozone Depletion and Global Warming. Proceeding of Third International Conference on Energy Research and Development (ICERD-3), Nov. 21-23, 2005 1:39-46

Global Pesticide Campaigner, 2004. Vol. 14, No. 2. www.panna.org/resources/gpc/gpc_200408.14.2.pdf

Gale, C.R., martyn, C.N., Winter, P.D. and Cooper, C., 1995. Vitamin C and Risk of Death from Stroke and Coronary Heart Disease in Cohort of Elderly People. *BMJ* 310:1563-1566

Gilbert, B. and Zhang, H. 2003. Nanoparticles Change Crystal Structure When They Get Wet. Research Shows. *Nature*. August 27.

Guo, R. et al. 2004. Thermal and Chemical Stabilities of Arsenic in Three Chinese Coals. *Fuel Processing Technology,* 85(8-10): 903-912.

Gerpen, J.V., Pruszko, R., Shanks, B., Clements, D., and Knothe, G., 2004. Biodiesel Analytical methods. National Renewable Energy Laboratory. Operated for the U.S. Department of Energy Office of Energy Efficiency and Renewable Energy. NREL/SR-510-36240

Houghton, J.T., Meira Filho, L.G., Callander, B.A., Harris, N., Kattenberg, A. and Maskell, K. Climate Change 2001: The Scientific Basis. Technical Summary. Cambridge University Press, Cambridge, UK.

Houghton, J., 2004. Global Warming: The Complete Briefing. Third Edition. Cambridge University Press, Cambridge, UK.

Hughes, L. and Scott, S., 1997. Canadian Greenhouse Gas Emissions:1990-2000. Energy Conversion and Management 38 (3)

Hossain M.E. and Islam, M.R., 2006. Fluid Properties with Memory – A Critical Review and Some Additions. 36th International Conference on Computers and Industrial Engineering (ICCIE) in Taipei, Taiwan, R.O.C., June 20-23, 2006.

IEO, 2005. Energy Information Administration / International Energy Outlook. Energy-Related Carbon Dioxide Emissions. www.eia.doe.gov/oiaf/ieo/pdf/emissions.pdf <accessed on May 29, 2006>.

IEA, 2005. Energy Information Administration / International Energy Outlook. Worldwide CO2 Emission from Fossil fuel Consumption and Flaring. www.eia.doe.gov/pub/international/iealf/tableh1co2.xls <accessed on May 30, 3006)

IPCC, 2001a. Climate Change 2001: The Scientific Basis. Houghton J.T., Ding, Y., Griggs, D.J., Noguer, M., Van der Linden, P.J., . Dai, X., Maskell, K. and C.A. Johnson, (eds), Cambridge University Press, Cambridge, UK, 881 pp.

IPCC, 2001b. Climate Change 2001: Report of the working group-I: The Scientific Basis. The Projections of Future Climate Change, Chapter 9.

Islam, M.R. 2003. Revolution in Education, EECRG, Nova Scotia, Canada (ISBN 0-9733656-1-7) 553 pp.

Islam, M.R., 2004. Unraveling the Mysteries of Chaos and Change: Knowledge-Based Technology Development", *EEC Innovation*, vol. 2, no. 2 and 3, 45-87.

Islam, M.R., 2005. Knowledge-Based Technologies for the Information Age, JICEC05-Keynote speech, Jordan International Chemical Engineering Conference V, 12-14 September 2005, Amman, Jordan

Islam, M.R.,2006a. Computing for the Information Age. Keynote speech, proceedings of the 36th International Conference on Computer and Industrial Engineering, Taiwan, June 20-24, 2006.

Islam, M.R.,2006b. A Knowledge-Based Water and Waste-Water Management Model. International Conference on Management of Water, Wastewater and Environment: Challenges for the Developing Countries, September 13-15, 2006, Kathmandu, Nepal.

Islam, M.R., Zatzman, G.M. and Shapiro. R.2006. "The Energy Crunch: What More Lies Ahead", in Global Dialogue on Energy [No. 2 in a series], 3-4 April, Washington DC at the Centre for Strategic and International Studies.

Icenhower, M.W.2006. Earth's Dwindling Water Supply. The Real Truth. http://www.realtruth.org/articles/0403-edws.html

Josefson, D.2003. Vitamin Supplements Do not Reduce the Incidence of Cancer or Heart Diseases. *BMJ* 327:70

Kaoma, J. and Kasali, G.B., 1994. Efficiency and Emissions of Charcoal use in the Improved Mbuala Cookstoves. Published by the Stockholm Environment Institute in Collaboration with SIDA, ISBN:91 88116: 94 8.

Khan, M.I., 2006. Towards Sustainability in Offshore Oil and Gas Operations, Ph.D. Dissertation, Dalhousie University, Halifax, Canada, 442 pp.

Khan, M.I. and Islam, M.R., 2006a. Handbook of Sustainable Oil and Gas Operations Management, Gulf Publishing Company, USA: in press.

Khan, M.I. and Islam, M.R., 2006b. True Sustainability in Technological Development and Natural Resources Management. Nova Science Publishers, New York, USA: in press.

Khan, M.I, and Islam, M.R., 2005a. A Novel Sustainability Criterion as Applied in Developing Technologies and Management Tools. Sustainable Planning 2005, 12 - 14 September 2005, Bologna, Italy.

Khan, M.I., and Islam, M.R., 2005b. Assessing the Sustainability of Technological Developments: An Alternative Approach of Selecting Indicators in the Case of Offshore Operations, ASME International Mechanical Engineering Congress and Exposition (IMECE), Orlando, Florida, USA, November.

Khan, M.M., Prior, D. and Islam, M.R., 2006a. A Novel, Sustainable Combined Heating/Cooling/Refrigeration System. *J. Nat. Sci. and Sust. Tech.* vol. 1, no. 1: in press.

Khan, M.I., Chhetri, A.B. and Islam, M.R. 2006b. Achieving True Technological Sustainability: Pathway Analysis of a Sustainable and an Unsustainable Product. *J. Nat. Sci. and Sust.Tech.* Submitted.

Khan, M.I, Zatzman, G. and Islam, M.R., 2005a. New Sustainability Criterion: Development of Single Sustainability Criterion as Applied in Developing Technologies. Jordan International Chemical Engineering Conference V, Paper No.: JICEC05-BMC-3-12, Amman, Jordan, 12 - 14 September 2005.

Khan, M.I., Chhetri, A.B. and Islam, M.R., 2006c. Community-Based energy Model: A Novel Approach in Developing Sustainable Energy. Energy Sources: in press.

Khan, M.M., Prior, D. and Islam, M.R., 2005b. Jordan International Chemical Engineering Conference V, 12-14 September, Amman, Jordan.

Khilyuk, L.F. and Chilingar, G.V. 2003. Global Warming: Are We Confusing Cause and Effect? *Energy Sources* 25:357-370

Khilyuk, L.F. and Chilingar, G.V. 2004. Global Warming and Long Term Climatic Changes: A Progress Report. *Environmental Geology* 46(6-7):970-979.

Khilyuk, L.F. and Chilingar, G.V. 2006. On Global Forces of Nature Driving the Earth's Climate. Are Humans Involved? *Environmental Geology.* Published on line.

Khilyuk, L.F., Katz, S.A., Chilingarian, G.V. and Aminzadeh, F. 1994. Global Warming: Are We Confusing Cause and Effect? *Energy Sources* 25:357-370

Kondratyev, K.Y.A. and Cracknell, A. P. 1998. Observing Global Climate Change. Taylor and Francis. ISBN- 0748401245 pp:544

Kyoto Protocol, 1997. Conference of the Parties Third Session Kyoto, 1-10 December 1997. Kyoto Protocol to the United Nations Framework Convention on Climate Change.

Lakhal, S. Y., and H'Mida, S. (2003). A Gap Analysis for Green Supply Chain Benchmarking. In "32th International Conference on Computers and Industrial Engineering", Vol. Vol. 1, pp. 1: 44-49, Ireland, August 11- 13th, 2003

Lindzen, R.S., 2002. Global Warming: The Origin and Nature of the Alleged Scientific Consensus.Regulation:The Cato Review of Business and Government. http://eaps.mit.edu/faculty/lindzen/153_Regulation.pdf

Lindzen, R.S., 2006. Climate Fear. The Opinion Journal, April, 12, 2006. www.opinionjournal.com/ extra/?id=110008220 (Accessed on June, 30, 2006).

Martinot, E., 2005. Renewable Energy Policy Network for the 21st Century. Global Renewables Status Report Prepared for the REN 21 Network by the Worldwatch institute.

NASA, 1999. Biomass Burning and Global Change. http://asd www.larc.nasa.gov/ biomass_burn/biomass_burn.html

NOAA, 2005a. Trends in Atmospheric Carbon Dioxide. NOAA-ESRL Global Monitoring Division. www.cmdl.noaa.gov/ccgg/trends/(accessed on June 04, 2006)

NOAA, 2005b. Greenhouse Gases, Global Monitoring Division, Earth System Research Laboratory, National Oceanic and Atmospheric Administration, USA

Pershing, J. and Cedric, P., 2002. Promises and Limits of Financial Assistance and theClean Development Mechanism. Beyond Kyoto: Energy Dynamics and Climate Stabilization. Paris: International Energy Agency, 94-98

Rahman, M.H., Wasiuddin, N.M. and Islam, M. R. 2004. Experimental and Numerical Modeling Studies of Arsenic Removal with Wood Ash from Aqueous Streams. *The Canadian Journal of Chemical Engineering* 82:968-977

Rahman, M.S. and Islam, M.R., 2006. Environmentally friendly Alkaline Solution for Enhanced Oil Recovery. *Nat.Sci.and Sust.Tech.* submitted.

Service, R.F., 2005. Is it time to shoot for the sun? *Science*, Vol. 309: 549-551.

Shastri, C.M., Sangeetha, G. and Ravindranath, N.H. 2002. Dissemination of efficient ASTRA stove: case study of a successful entrepreneur in Sirsi, India. Energy for Sustainable Development.Volume VI., No. 2

Sudan, B.J.L., 1997. Total abrogation of facial seborrhoeic dermatitis with extremely low-frequency (1-1.1 Hz) 'imprinted' water is not allergen or hapten dependent: a new visible model for homoeopathy. Medical Hypotheses 48: 477-479.

The Epoch Times, 2006. *Potato Farms a Hot Bed for Cancer.* March 24-30, 2006. www.theepochtimes.ca.

The Globe and Mail, 2006. Toxic shock: Canada's Chemical reaction. May 27, Saturday, 2006.

Thomas P. and Nowak, M.A., 2006. Climate Change: All in the Game. *Nature* (441) June 1, 2006

Tschulakow, A.V., Yan, Y. and Klimek, W., 2005. A New Approach to the Memory of Water. *Homeopathy* 94 (4): 241-247

The New York Time, 2006. Citing Security, Plants Use Safer Chemicals. April, 25, 2006.

Vafaei, S., 2006. A High Efficiency Jet Engine. Master's Degree Thesis, Dalhousie University, Department of Civil and Resources Engineering.

WEC, 2006.The World Energy Council: How to Avoid a Billion Tones of CO2 Emission. http://www.worldenergy.org/wec-geis/default.asp <accessed on May 30, 2006>

Wasiuddin, N.M., Tango, M. and Islam, M.R. 2002. A Novel Method for Arsenic Removal at Low Concentrations. *Energy Sources* 24, 1031–1041

Wise Uranium Project, 2005. Uranium Radiation Properties. www.wise-uranium.org/rup.html (accessed on March 19, 2006)

York, M., 2003. One Spoonful of Bee Pollen Each Day, and You, Too, Might Make It to 113. The New York Times, December, 2003 <accessed on June 06, 2006>.

Zatzman, G.M. and Islam, M.R., 2006, Economics of Intangibles, Nova Science Publishers, New York: in press.

In: Nature Science and Sustainable Technology
Editor: M. R. Islam, pp. 119-135
ISBN: 978-1-60456-009-1
© 2008 Nova Science Publishers, Inc.

Chapter 6

THE ADOMIAN DECOMPOSITION METHOD ON SOLUTIONS OF NON-LINEAR PARTIAL DIFFERENTIAL EQUATIONS

S. H. Mousavizadegan[1], S. Mustafiz[*1] and M. Rahman[2]

[1] Department of Civil and Resource Engineering, Dalhousie University, Canada
[2] Dept. of Engineering Mathematics, Dalhousie University, Canada

ABSTRACT

The Adomian decomposition method (ADM) is applied to solve nonlinear differential equations of various types. The ADM solution is an infinite series that is constructed using the initial and boundary conditions and the forcing function. The numerical result is approximated by truncation of the series solution to a finite number. The effect of this truncation on the final result is investigated in this paper. The accuracy of the ADM significantly depends on the number of series solution terms. The solution is good for a very small value of the independent variables. The ADM solution is susceptible to severe instability with increasing the spatial or time variables.

Keywords: Adomian decomposition method; nonlinear PDE; power polynomial; series solution.

INTRODUCTION

The methods of the solution of the nonlinear partial differential equations (PDE) and integral equations (IE) may be categorized as exact, approximate and numerical methods. The exact solutions are constructed by using analytical procedures. In the approximate methods, the solutions are obtained in the form of functions which are close to the exact solution. The

---

[*] Corresponding author: Email: mustafiz@dal.ca

Adomian decomposition method is an approximate scheme in which the solution is obtained recursively using the forcing function and the initial and boundary conditions. This method was first proposed by a North American physicist, G. Adomian (1923-1996) and well addressed in Adomian (1984, 1989 and 1994). The ADM solution is obtained in a series form while the nonlinear term is decomposed into a series in which the terms are calculated recursively using Adomian polynomials.

There are several published papers on the solution of the linear and nonlinear partial differential equation using the Adomian decomposition method. Recently, Biazar and Islam (2004) have used ADM to solve the linear wave equation in 3-dimensional space. They indicate that ADM is very sensitive to initial and boundary conditions and do not function if the initial and boundary conditions are constant. The nonlinear dispersive KdV equation with initial profile is solved by Wazwaz (2001) using ADM. The approximate solutions are arranged in power polynomials that are applied to find out the closed form solutions for different test examples. The truncated ADM solution for the two solitons solution are reported for a very small time interval and does not match with the closed form solution of the problem. Deeba and Khuri (1996) have applied ADM for solving the nonlinear Klein-Gordon (KG) Equation. The tabulated errors for a nonlinear test example show that the accuracy of the ADM solution is impaired with increasing time intervals.

We investigate the influence of the truncation of the ADM series solution on the accuracy and the stability of the results in this article. The ADM is applied to solve the nonlinear Burger's equation with various initial and boundary conditions, the nonlinear KdV equation with one and two solitons solution and the KG-equation of different types. The formulation of the Adomian decomposition method is presented for the general case of functional equations and then extended to each of these PDEs. Several examples for each type of these PDEs are taken into account and the ADM solutions are obtained for different number of series solution terms. The approximate solutions are compared with the closed form solutions that are obtained from the published literatures. All numerical results are obtained using MATLAB.

ADOMIAN DECOMPOSITION METHOD

Let us consider the non-linear functional equation in the form

$$u(\mathbf{x},t) = f(\mathbf{x},t) + L[u(\mathbf{x},t)] + N[u(\mathbf{x},t)], \tag{1}$$

where, L and N represent known linear and nonlinear operators respectively and $f(x,t)$ is a given function. The problem is to determine the solution $u(x,t)$ of (1). This solution may be expressed in a series form as $u(x,t) = \sum_{n=0}^{\infty} u_n$. The functional equation (1) may be written in the following form using the ADM.

$$\sum_{n=0}^{\infty} u_n = f(\mathbf{x},t) + \sum_{n=0}^{\infty} B_n + \sum_{n=0}^{\infty} A_n \tag{2}$$

The term denoted by B_n is associated with the linear operator L and A_n is the Adomian polynomial element that is given by

$$A_n = \frac{1}{n!} \frac{d^n}{d\lambda^n} \left[N\left(\sum_{i=0}^{\infty} \lambda^i u(\mathbf{x},t)_i \right) \right]_{\lambda=0}. \tag{3}$$

where, λ is a parameter introduced for convenience sake. The elements of the series solution for $u(x,t)$ are obtained recursively by

$$u_0(\mathbf{x},t) = f(\mathbf{x},t), \quad u_1(\mathbf{x},t) = B_0 + A_0, \quad \cdots \quad u_k(\mathbf{x},t) = B_{k-1} + A_{k-1} \quad \cdots . \tag{4}$$

By this arrangement, the linear and nonlinear part of the functional equation (1) is replaced by a known function in each step using the recursive equations (4).

APPLICATION OF THE METHOD

The method is applied to three partial differential equations with various initial and boundary conditions. The PDEs that are taken into account are:

- The Burgers equation that is the simplest model for the one-dimensional turbulence in fluid flow. This equation can be also applied to describe many physical events such as sound-wave in a viscous medium, waves in fluid-filled viscous elastic tubes, etc. (Debnath, 2005).

$$\begin{aligned} &u_t + uu_x = \nu u_{xx} \quad \text{and} \\ &u(x,0) = u_0(x), \quad \text{where} \quad a < x < b \\ &u(a,t) = C_1(t), \quad u(b,t) = C_2(t), \quad \text{where} \quad t > 0 \end{aligned} \tag{5}$$

where, ν is the kinematic viscosity

- The Korteweg-de Vries (KdV) equation that describes the motion of nonlinear long wave in shallow water under gravity

$$\begin{aligned} &u_t - 6uu_x + \nu u_{xxx} = 0 \quad \text{and} \\ &u(x,0) = u_0(x), \quad \text{where} \quad |x| < \infty \end{aligned} \tag{6}$$

- The Klein-Gordon (KG) equation in the from

$$u_{tt} - \sum_{j=1}^{m}\left\{\partial_{x_j}\left[f_j(\mathbf{x})u_{x_j}\right]\right\} + g(u(\mathbf{x},t)) = h(\mathbf{x},t)$$
$$u(\mathbf{x},0) = C_1(\mathbf{x}), \quad u_t(\mathbf{x},0) = C_2(\mathbf{x}), \quad \text{where} \quad |\mathbf{x}| < \infty \tag{7}$$

where, $h(x,t)$ is the forcing function.

The Burgers Equation

The Burgers equation (5) may be integrated with respect to time variable and written as

$$u(x,t) = u(x,0) + \int_0^t \nu u_{xx} dt - \int_0^t u u_x dt. \tag{8}$$

Using the ADM, (8) can be written as (2), where

$$f(x,t) = u(x,0)$$
$$B_n = \int_0^t \nu(u_n)_{xx} dt$$
$$A_n = \int_0^t \left\{ \frac{1}{n!} \frac{d^n}{d\lambda^n}\left\{\left[\sum_{i=0}^{\infty}(\lambda^i u_i)\right]\left[\sum_{i=0}^{\infty}\left(\lambda^i (u_i)_x\right)\right]\right\}\right\}_{\lambda=0} dt, \tag{9}$$

and the elements of the series solution are obtained according to (4).

$$u_0 = u(x,0)$$
$$u_1 = \int_0^t \left[\nu(u_0)_{xx} - u_0(u_0)_x\right] dt$$
$$u_2 = \int_0^t \left[\nu(u_1)_{xx} - \left[u_0(u_1)_x + u_1(u_0)_x\right]\right] dt$$
$$\cdots \quad \cdots \quad \cdots \quad \cdots$$
$$u_n = \int_0^t \left[\nu(u_{n-1})_{xx} - \left[u_0(u_{n-1})_x + u_1(u_{n-2})_x + \cdots + u_{n-1}(u_0)_x\right]\right] dt$$
$$\cdots \quad \cdots \quad \cdots \quad \cdots \tag{10}$$

It can be observed from (10) that if the initial condition is constant, the ADM is not any more applicable for this type of PDE and the solution will be the same as initial value.

The Korteweg-de Vries (KdV) Equation

The KdV equation (6) is integrated with respect to time variable and written as

$$u = u(x,0) + 6\int_0^t uu_x dt - \int_0^t u_{xxx} dt. \tag{11}$$

The solution can be constructed using ADM in series form as (2).

$$u_0 = u(x,0), \quad u_1 = 6A_0 - B_0, \quad \cdots, u_n = 6A_{n-1} - B_{n-1}, \tag{12}$$

The components of B_n series are computed by $B_n = \int_0^t (u_n - 1)_{xxx} dt$ and the A_n series terms are found the same as given in (9) for the Burgers equation. The series solution terms are computed recursively by

$$u_0 = u(x,0)$$
$$u_1 = \int_0^t \left[6 u_0(u_0)_x - (u_0)_{xxx} \right] dt$$
$$u_2 = \int_0^t \left[6[u_0(u_1)_x + u_1(u_0)_x] - (u_1)_{xxx} \right] dt$$
$$\cdots \cdots \cdots \cdots$$
$$u_n = \int_0^t \left[6[u_0(u_{n-1})_x + u_1(u_{n-2})_x + \cdots + u_{n-1}(u_0)_x] - (u_{n-1})_{xxx} \right] dt$$
$$\cdots \cdots \cdots \cdots . \tag{13}$$

If the initial condition is constant, the ADM is not any more functional for this type of PDE.

The Klein-Gordon (KG) Equation

The Klein-Gordon (KG) equation (7) may be written in the following form by integrating in respect to time variable.

$$u = u(\mathbf{x},0) + u_t(\mathbf{x},0)t + \int_0^t\int_0^t h(\mathbf{x},t)\,(dt)^2 + \int_0^t\int_0^t \sum_{j=1}^m \left\{ \partial_{x_j}\left[f_j(\mathbf{x})u_{x_j}\right] \right\} (dt)^2$$
$$- \int_0^t\int_0^t g(u)\,(dt)^2 \tag{14}$$

The function $g(u)$ may be written as $g(u) = Lg(u) + Ng(u)$, where $Lg(u)$ is the linear part and $Ng(u)$ is the nonlinear part. Using ADM, the solution is approximated by a series solution as $u(x,t) = \sum_{n=0}^{\infty} u_n$ and (14) is expressed in the form of (2) where

$$f(\mathbf{x},t) = u(\mathbf{x},0) + u_t(\mathbf{x},0)t + \int_0^t\!\!\int_0^t h(\mathbf{x},t)\,(dt)^2$$

$$B_n = \int_0^t\!\!\int_0^t \sum_{j=1}^m \left\{\partial_{x_j}\!\left[f_j(\mathbf{x})(u_{n-1})_{x_j}\right]\right\}(dt)^2 - \int_0^t\!\!\int_0^t Lg(u_{n-1})\,(dt)^2$$

$$A_n = -\int_0^t\!\!\int_0^t \frac{1}{n!}\frac{d^n}{d\lambda^n}\left[Ng\!\left(\sum_{i=0}^\infty \lambda^i u(\mathbf{x},t)_i\right)\right]_{\lambda=0}(dt)^2. \tag{15}$$

The series solution may be constructed according to (4). The function $f(x,t)$ may be composed of several parts as given in (15). For convenience of computation, some part of the the whole of $f(x,t)$ may be added to the other series solution terms instead of u_0.

IMPLEMENTATION OF ADM

The Adomian decomposition method (ADM) is applied to several test examples of the nonlinear PDEs that are explained in previous section. The object is to study the effect of truncation of the series solution on the accuracy and the stability of results.

The Burgers Equation

The ADM is implemented to the Burgers equation with three different conditions.

Example 1
In the first example, the initial condition is

$$u(x,0) = \frac{x}{\nu} + 1, \quad \text{and} \quad |x| < \infty. \tag{16}$$

Using ADM, a series solution can be constructed according to (10) as

$$u_0 = \frac{x}{\nu} + 1$$

$$u_1 = -\frac{t}{\nu}(\frac{x}{\nu} + 1)$$

$$u_2 = \frac{t^2}{\nu^2}(\frac{x}{\nu} + 1)$$

$$u_3 = -\frac{t^3}{\nu^3}(\frac{x}{\nu} + 1)$$

$$u_4 = \frac{t^4}{\nu^4}(\frac{x}{\nu} + 1)$$

$$\ldots \quad \ldots \quad \ldots \quad \ldots$$

$$u_n = (-1)^n \frac{t^n}{\nu^n}(\frac{x}{\nu} + 1). \tag{17}$$

The final solution may be written in the form of

$$u = (\frac{x}{\nu} + 1)(1 - \frac{t}{\nu} + \frac{t^2}{\nu^2} - \frac{t^3}{\nu^3} + \cdots + (-1)^n \frac{t^n}{\nu^n} + \cdots) = (\frac{x}{\nu} + 1)\sum_{n=0}^{\infty}(-1)^n \frac{t^n}{\nu^n}. \tag{18}$$

The closed form solution is in the form $u = (x/\nu + 1)(1 + t/\nu)^{-1}$ for this initial value problem.

The solution are illustrated in Fig. 1 as a function of time at x = 1 for ν = 1. The ADM solution is shown for different truncation number M. The closed form solution is also shown to provide a comparison among them. The ADM solutions even with small value of truncation number are in excellent agreement with the exact solution for small values of t. However, the accuracy of the approximate solution decreases with increasing the time interval. The inception point that the approximate solution deviates from the exact solution depends on the truncation number M. The diagrams in Fig. 1 show that the ADM solution is applicable for a limited time interval.

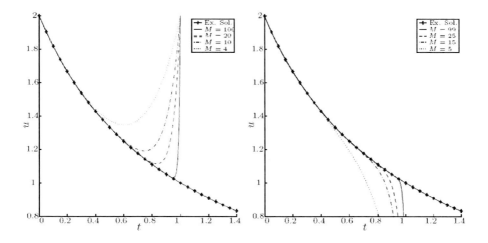

Figure 1. The effect of truncation of the ADM series solution for the first example of the Burgers equation, $\nu = 1$, x = 1, Left: M is even number Right: M is odd number

Example 2
It is an initial value problem with the initial condition as

$$u(x,0) = 2(1+e^{\frac{x}{\nu}})^{-1}, \quad \text{and} \quad |x| < \infty. \tag{19}$$

Some of the first elements of the series solution according to ADM are

$$u_0 = 2(1+e^{x/\nu})^{-1}$$
$$u_1 = 2(t/\nu)\, e^{x/\nu}(1+e^{x/\nu})^{-2}$$
$$u_2 = (t/\nu)^2\, e^{x/\nu}(e^{x/\nu}-1)(1+e^{x/\nu})^{-3}$$
$$u_3 = \frac{1}{3}(t/\nu)^3\, e^{x/\nu}(1-4e^{x/\nu}+e^{2(x/\nu)})(1+e^{x/\nu})^{-4}$$
$$u_4 = \frac{1}{12}(t/\nu)^4\, e^{x/\nu}(e^{x/\nu}-1)(e^{2(x/\nu)}-10e^{x/\nu}+1)(1+e^{x/\nu})^{-5}$$
$$\cdots \quad \cdots \quad \cdots \quad \cdots$$
$$u_n = \frac{2}{n!}(t)^n \frac{d^n}{dt^n}\left(\frac{2}{1+e^{(x-t)/\nu}}\right)_{t=0}. \tag{20}$$

The closed form solution can be written as $u = \dfrac{2}{1+e^{(x-t)/\nu}}$ according to the Taylor series expansion as $n \to \infty$.

The effects of the truncation of the series solution to a finite number M are illustrated in Fig. 2 for this test example. The computations are carried out at $x = 1$ for $\nu = 1$ at different time intervals. The same trend as the first example is observed. The accuracy of the ADM solution significantly depends on the truncation number M. The other difficulty is the instability of the approximate solution with increasing M. The ADM solutions deviate from the exact solution at a small value of t when $M = 99, 100$ and are unstable after that time. This is due to the inherent properties of the power polynomials that are susceptible to the severe instability with increasing order.

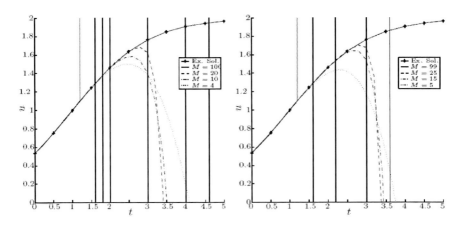

Figure 2. The effect of truncation of the ADM series solution for the second example of the Burgers equation, $\nu = 1$, x = 1, Left: M is even number Right: M is odd number

Example 3

The Adomian decomposition method is also implemented to an example that shows the decay of an arbitrary periodic initial disturbance. This can be taken into account as a theoretical model for free turbulence in a box. It is assumed that

$$u(x,0) = \sin \pi x \quad \text{and} \quad 0 \leq x \leq 1$$
$$u(0,t) = u(1,t) = 0. \tag{21}$$

The analytical solution for this problem is given by Cole (1951) as

$$u(x,t) = \frac{4\pi\nu \sum_{n=1}^{\infty} n I_n(\frac{1}{2\pi\nu}) \sin n\pi x e^{-n^2\pi^2\nu t}}{I_0(\frac{1}{2\pi\nu}) + 2\sum_{n=1}^{\infty} n I_n(\frac{1}{2\pi\nu}) \cos n\pi x e^{-n^2\pi^2\nu t}} \tag{22}$$

Where I_n is the modified Bessel function of first kind and order n.

Substituting (21) into (13), the elements of the series solution are found for this example as

$$u_0 = \sin \pi x$$
$$u_1 = -\pi t \left(\nu\pi \sin \pi x + \frac{1}{2} \sin 2\pi x \right)$$
$$u_2 = \frac{\pi^2 t^2}{2} \left[\left(\nu^2 \pi^2 - \frac{1}{4} \right) \sin \pi x + 3\nu\pi \sin 2\pi x + \frac{3}{4} \sin 3\pi x \right]$$
$$u_3 = -\frac{\pi^3 t^3}{6} \left[\nu\pi \left(\nu^2 \pi^2 - \frac{9}{4} \right) \sin \pi x + (14\nu^2\pi^2 - 1)\sin 2\pi x + \frac{51}{4} \nu\pi \sin 3\pi x + 2\sin 4\pi x \right]$$
$$u_4 = \frac{\pi^4 t^4}{24} \left[\left(\nu^4\pi^4 - \frac{245}{2}\nu^2\pi^2 - 7 \right) \sin \pi x + (60\nu^3\pi^3 - 22)\sin 2\pi x + \left(\frac{29}{2}\nu^2\pi^2 + \frac{147}{16} \right) \sin 3\pi x + (60\nu\pi + 11)\sin 4\pi x + \frac{11}{16}\sin 5\pi x \right]$$
$$\ldots \ldots \ldots \ldots \tag{23}$$

The ADM solution may be constructed as $u(x,t) = \sum_{n=0}^{\infty} u_n$ where M is the truncation number.

The computations are carried out at different time intervals as a function of x when $M = 20$ for $\nu = 0.01$. The approximate solutions are depicted in Fig. 3 along with the analytical solution. The ADM solutions comply with the analytical solution for a limited time. The ADM solutions for $t > 0.25$ starts to oscillate and for $t > 0.4$, they become completely unstable.

If the nonlinear part of the Burgers equation is omitted, the equation reduces to a one dimensional heat equation, $u_t = \nu u_{xx}$. If the same initial and boundary conditions as (21) are used, the ADM solution can be constructed as

$$u = \sin\pi x \left[1 - \nu\pi^2 t + \frac{1}{2!}(\nu\pi^2 t)^2 - \frac{1}{3!}(\nu\pi^2 t)^3 + \frac{1}{4!}(\nu\pi^2 t)^4 - \cdots + (-1)^n \frac{1}{n!}(\nu\pi^2 t)^n\right]$$

$$= \sin\pi x \sum_{n=0}^{\infty}(-1)^n \frac{1}{n!}(\nu\pi^2 t)^2. \qquad (24)$$

If $n \to \infty$, the solution corresponds to the closed form of $u = \sin\pi x \exp(-\nu\pi^2 t)$.

In order to see the effect of truncation of the series, The solution is examined for different number of the series elements in Fig. 4. The solutions are the same as the closed form solution up to a certain time and after that the solution will diverges. If M is reduced the deviation of the ADM solution from the closed form starts earlier.

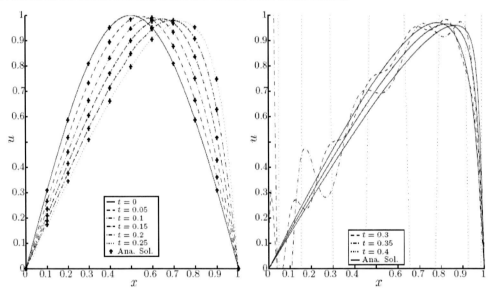

Figure 3. The comparison of the ADM and closed form solution of the third example of the Burgers equation, $\nu = 0.01$.

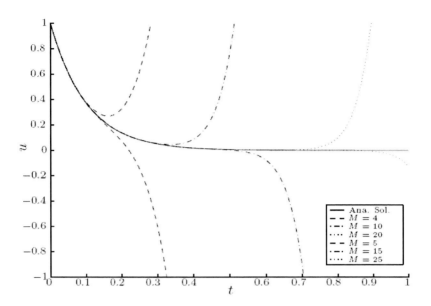

Figure 4. Deviation of the ADM solution from the close form solution due to the truncation of the series solution at different number for the linear heat equation at $x = 0.5$ and $v = 1$

The KdV Equation

Example 1

The first example is a single-soliton solution for a sech^2 potential. If the KdV equation is subjected to the initial condition of $u(x,0) = -2\sec h^2 x$, $|x| < \infty$, the ADM series solution can be constructed according to (13) as following forms.

$$u_0 = -2\,\text{sech}^2 x$$
$$u_1 = -16\,t\,\sinh x\,\text{sech}^3 x$$
$$u_2 = -32\,t^2\,(2\cosh^2 x - 3)\,\text{sech}^4 x$$
$$u_3 = -\frac{512}{3}\,t^3\,\sinh x(\cosh^2 x - 3)\,\text{sech}^5 x$$
$$u_4 = -\frac{512}{3}\,t^4\,(2\cosh^4 x - 15\cosh^2 x + 15)\,\text{sech}^6 x$$
$$\ldots \quad \ldots \quad \ldots \quad \ldots \tag{25}$$

The closed form solution of this problem is $u(x,t) = -2\sec h^2(x - 4t)$.

The computational results of u using the ADM are shown in Fig. 5 as a function of spatial variable at different elapsing time. The analytical solution is also presented. The ADM solution is obtained with $M = 20$. The ADM results are in excellent agreement with the analytical solution for a very small elapsing time ($t = 0.3$). The ADM solution becomes unstable for a elapsing time of $t > 0.3$.

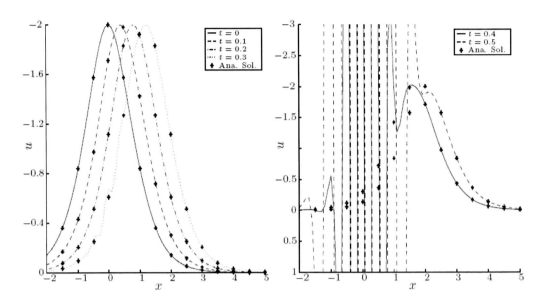

Figure 5. The single-soliton solution of the KdV equation at different elapsing time, $M = 20$.

Example 2
It is a two-soliton solution for a sech^2 potential. The KdV equation is subjected to the initial condition of $u(x,0) = -6\,\text{sech}^2 x$, $|x| < \infty$. The ADM series solution is constructed according to (13) for the first four terms

$$u_0 = -6\,\text{sech}^2 x$$
$$u_1 = -48\,t\,\sinh x(\cosh^2 x + 6)\,\text{sech}^5 x$$
$$u_2 = -96\,t^2\,(2\cosh^6 x + 117\cosh^4 x - 96\cosh^2 x - 63)\,\text{sech}^8 x$$
$$u_3 = -256\,t^3\,\sinh x(2\cosh^8 x + 1002\cosh^6 x - 594\cosh^4 x$$
$$-1728\cosh^2 x - 405)\,\text{sech}^{11} x. \qquad (26)$$

The two-soliton solution of the KdV equation using ADM is shown in Fig. 6 along with the closed form solution at different time interval. Moreover, the closed form solution of this problem is given in Rahman (1995)

$$u(x,t) = -12\frac{3 + 4\cosh(8t - 2x) + \cosh(64t - 4x)}{\bigl(\cosh(36t - 3x) + 3\cosh(28t - x)\bigr)^2}. \qquad (27)$$

In this case, the ADM solution are obtained with $M = 18$. Once again, the ADM results are found to be in excellent agreement with the analytical solution for a very small elapsing time. The solution with ADM becomes unstable for $t > 0.05$.

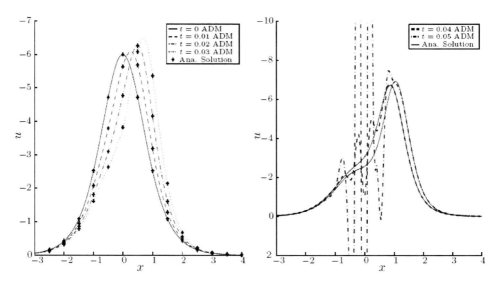

Figure 6. The two-soliton solution of the KdV equation at different elapsing time, $M = 18$

The KG Equation

Example 1
The first test example is a linear nonhomogeneous KG-equation in the form

$$u_{tt} - u_{xx} + 2u = -\sin x \sin t$$
$$u(x,0) = 0 \quad, \quad u_t(x,0) = \sin x, \quad \text{and} \quad u(0,t) = u(\pi,t) = 0. \tag{28}$$

The ADM series solution according to (15) may be constructed as

$$u_0 = \sin x(2\sin t - t)$$
$$u_1 = \sin x(-2\sin t + 2t - 1/3!\, t^3)$$
$$u_2 = \sin x(2\sin t - 2t + 2/3!\, t^3 - 1/5!\, t^5)$$
$$u_3 = \sin x(-2\sin t + 2t - 2/3!\, t^3 + 2/5!\, t^5 + 1/7!\, t^7)$$
$$u_4 = \sin x(2\sin t - 2t + 2/3!\, t^3 - 2/5!\, t^5 - 2/7!\, t^7 + 1/9!\, t^9)$$
$$\cdots \quad \cdots \quad \cdots \quad \cdots \,. \tag{29}$$

The solution of the problem can be written in a series form, which is as follows

$$u(x,t) = \sin x \left[2\sin t - \left(t - \frac{1}{3!}t^3 + \frac{1}{5!}t^5 - \frac{1}{7!}t^7 + \frac{1}{9!}t^9 - \frac{1}{11!}t^{11} + \frac{1}{13!}t^{13} - \cdots \right) \right]$$
$$= \sin x \left[2\sin t - \sum_{k=0}^{2M} \frac{(-1)^{k/2}}{(k+1)!} t^{k+1} \right] \quad \text{and} \quad M = 1,2,3,\cdots. \tag{30}$$

The closed form solution of Example 1 is $u(x,t) = \sin x \sin t$ as $M \to \infty$.

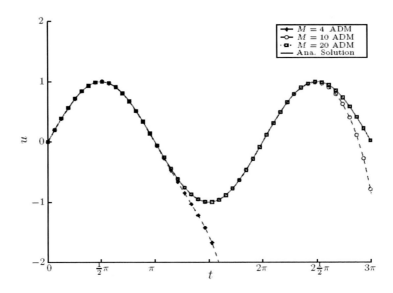

Figure 7. The effect of the truncation of the ADM series solution on a linear non-homogeneous KG-equation

The effect of the truncation of the series solution on the accuracy of the result is shown in Fig. 7. The solutions are obtained as a function of time for $x = \pi/2$ using different truncation number M. When $M = 20$, the ADM results are in excellent agreement with the closed form solution in the time range as shown in the Fig. 7. The ADM solution shows a very good agreement with the closed form solution even for $M = 4$. However, the accuracy of the results are impaired with increasing time.

Example 2
This example is a Sine-Gordon equation in the form

$$u_{tt} - u_{xx} + \sin u = 0$$

$$u(x,0) = 4\tan^{-1}(e^{2x}) \, , \quad u_t(x,0) = \frac{-4\sqrt{3}\ e^{2x}}{1+e^{4x}} \, , \quad \text{and} \quad |x| < \infty. \tag{31}$$

The computation of B_n terms, which are associated to the linear operators, is straight forward. The A_n terms due to the nonlinear sine-operator are obtained recursively by

$$A_0 = -\int_0^t\!\!\int_0^t \sin u_0 (dt)^2$$

$$A_1 = -\int_0^t\!\!\int_0^t u_1 \cos u_0 (dt)^2$$

$$A_2 = -\int_0^t\!\!\int_0^t \left(u_2 \cos u_0 - \frac{u_1^2}{2!}\sin u_0\right)(dt)^2$$

$$A_3 = -\int_0^t\!\!\int_0^t \left(u_3 \cos u_0 - u_2 u_1 \sin u_0 - \frac{u_1^3}{3!}\cos u_0\right)(dt)^2$$

$$A_4 = -\int_0^t\!\!\int_0^t \left(u_4 \cos u_0 - u_3 u_1 \sin u_0 - \frac{1}{2}u_2^2 \sin u_0 - \frac{1}{2}u_2 u_1^2 \cos u_0 + \frac{u_1^4}{4!}\sin u_0\right)(dt)^2$$

$$\ldots \quad \ldots \quad \ldots \quad \ldots . \tag{32}$$

If the regular form of the ADM is applied the computation will be very complicated. Therefore, the modified form of ADM is used according to Wazwaz (1999). It simplifies the computation and makes the ADM feasible for this type of nonlinear PDEs. It is assumed that $u_0 = 0$ and the other terms of the ADM series solution are constructed in the form

$$u_1 = 4\tan^{-1}(e^{2x}) - \frac{4\sqrt{3}\,e^{2x}}{1+e^{4x}}t$$

$$u_2 = -\frac{8\,e^{2x}\,t^2}{3(1+e^{4x})^3}\left(\sqrt{3}\,t\,e^{8x} + 3\,e^{8x} - 6\sqrt{3}\,t\,e^{4x} - 3 + \sqrt{3}\,t\right) - 2\,t^2\tan^{-1}(e^{2x}) + \frac{2\sqrt{3}\,t^3 e^{2x}}{3(1+e^{4x})}$$

$$u_3 = \frac{2e^{2x}t^4}{5(1+e^{4x})^5}\left(-\sqrt{3}\,t\,e^{16x} - 5\,e^{16x} + 100\sqrt{3}\,t\,e^{12x} + 150\,e^{12x} - 310\sqrt{3}\,t\,e^{8x} + 100\sqrt{3}\,t\,e^{4x}\right.$$
$$\left. - 150\,e^{4x} + 5 - \sqrt{3}\,t\right) - \frac{8\,e^{2x}}{3(1+e^{4x})^3}\left[\frac{t^5}{20}\left(-\sqrt{3}e^{8x} + 6\sqrt{3}e^{4x} - \sqrt{3}\right) + \frac{t^4}{12}\left(-3\,e^{8x} + 3\right)\right]$$
$$+ \frac{t^4}{6}\tan^{-1}(e^{2x}) - \frac{t^5\sqrt{3}\,e^{2x}}{30(1+e^{4x})}$$

$$u_4 = -\frac{e^{2x}\,t^6}{105(1+e^{4x})^7}\left(21\,e^{24x} + 3\sqrt{3}\,t\,e^{24x} - 8428\,e^{20x} - 3654\sqrt{3}\,t\,e^{20x} + 56013\sqrt{3}\,t\,e^{16x}\right.$$
$$+ 54761\,e^{16x} - 126420\sqrt{3}\,t\,e^{12x} + 56013\sqrt{3}\,t\,e^{8x} - 54761\,e^{8x} + 8428\,e^{4x} - 3654\sqrt{3}\,t\,e^{4x}$$
$$\left. - 21 + 3\sqrt{3}\,t\right) - \frac{2\,e^{2x}}{5(1+e^{4x})^5}\left[\frac{t^7}{42}\left(100\sqrt{3}\,e^{12x} - \sqrt{3}\,e^{16x} - 310\sqrt{3}\,e^{8x} + 100\sqrt{3}\,e^{4x} - \sqrt{3}\right)\right.$$
$$\left.+ \frac{t^6}{30}\left(-5e^{16x} - 150e^{4x} + 150e^{12x} + 5\right)\right] - \frac{8e^{2x}}{3(1+e^{4x})^3}\left[\frac{t^7}{840}\left(\sqrt{3}\,e^{8x} - 6\sqrt{3}\,e^{4x} + \sqrt{3}\right)\right.$$
$$\left.+ \frac{t^6}{120}\left(e^{8x} - 1\right)\right] + \frac{\sqrt{3}\,t^7\,e^{2x}}{1260(1+e^{4x})} + \frac{(1+e^{4x})^2}{5760\,e^{4x}}\left[4\tan^{-1}(e^{2x}) - \frac{4\sqrt{3}te^{2x}}{1+e^{4x}}\right]^5$$
$$- \frac{8(1+e^{4x})^2}{45\,e^{4x}}\left[\tan^{-1}(e^{2x})\right]^5 + \frac{8\sqrt{3}\,t(1+e^{4x})}{9\,e^{2x}}\left[\tan^{-1}(e^{2x})\right]^4 - \frac{t^6\tan^{-1}(e^{2x})}{180}$$

$$\ldots \quad \ldots \quad \ldots \quad \ldots . \tag{33}$$

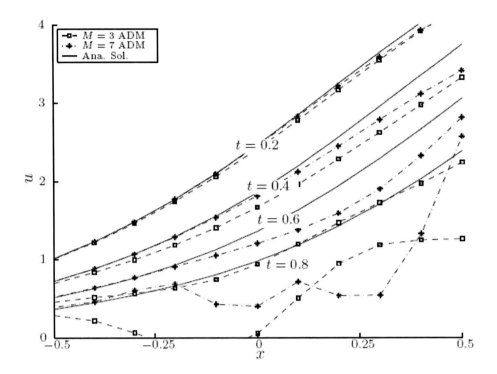

Figure 8. The effect of the truncation of the ADM series solution on the accuracy and stability of the results for a sine-Gordon equation

The approximate solution should be obtained by taking into account a finite number of series solution terms. The results are computed as a function of spatial variable at different time steps. The ADM solutions are obtained with $M = 4$ and 7 where M is the number of the series solution terms. The closed form solution of this problem is $u(x,t) = 4\tan^{-1}\left(e^{2x-\sqrt{3}t}\right)$, which is a solitary wave equation with the speed of $v = \sqrt{3}$ in the positive x-direction. This analytical solution is also illustrated in Fig. 8. The accuracy of the approximate solution is reasonable for small values of the time and spatial variables. The solution is susceptible to the severe instability as time and spatial coordinate increase. The solution can be improved with increasing the number of series solutions terms.

CONCLUSION

The Adomian decomposition method is applied to solve several partial differential equations of different types. A general formulation of the method is explained for nonlinear functional equations and then a detailed mathematical description of the method is presented for the nonlinear Burgers equation, the Korteweg-de Vries equation and the multi-dimensional Klein-Gordon equation. The method is implemented to several test examples of each type of these PDEs. The approximate solutions are found using the ADM with different number of series solution terms and compared with the closed form solution obtained from the published literatures. All the computations are carried out using the MATLAB software.

The Adomian decomposition method gives a stable solution with high degree of accuracy for a very small range of the independent variables in almost all test examples of these nonlinear PDEs. The accuracy of the approximate solution is impaired significantly and severe instability occure with increasing the values of the independent variables from certain points. The point at which the deviation of the approximate solution from the analytical results starts, depends on the truncation number of the ADM series solution. The point is also sensitive to the initial conditions. The accuracy of the ADM solution is improved and the deviation point is delayed with increasing the truncation number of the series solution. However, an increased in number of the series solution terms may result in an earlier occurrence of instability. The linear test examples show that the method may be more suitable for linear PDEs than the nonlinear PDEs.

ACKNOWLEDGEMENTS

The authors gratefully acknowledge the research grant provided through the Atlantic Innovation Fund (AIF). Mustafiz would also like to thank the Killam Foundation, Canada for its financial support.

REFERENCES

Adomian, G. and Adomian, G.E., 1984. A Global Method for Solution of Complex Systems, *Math. Model*, 5: 521-568.

Adomian, G., 1989. *Nonlinear Stochastic Systems Theory and Applications to Physics*, Kluwer Academic Publishers, MA.

Adomian, G., 1994. Solving Frontier Problems of Physics: The Decomposition Method, Kluwer Academic Publishers, Dordrecht.

Biazar, J. and Islam, R., 2004. Solution of Wave Equation by Adomian Decomposition Method and the Restrictions of the Method, *Applied Mathematics and Computation*, 149: 807-814.

Cole, J.D., 1951. On a Quasi-linear Parabolic Equations Occurring in Aerodynamics, *Quart. Appl. Math.*, 9: 225-236.

Debnath, L., 2005. Nonlinear Partial Differential Equations for Scientists and Engineers, Birkh¨auser Boston.

Deeba, E. Y. and Khuri, S. A., 1996. A Decomposition Method for Solving the Nonlinear Klein-Gordon Equation, *Journal of computational physics*, 124: 442-448.

Rahman, M., 1995. *Water Waves: Relating Modern Theory to Advanced Engineering Practice*, Oxford University Press Inc., New York.

Wazwaz, A.M., 1999. A Reliable Modification of Adomian Decomposition Method, *Appl. Math. Comput.*, 102: 77-86.

Wazwaz, A.M., 2001. Construction of Solitary Wave Solutions and Rational Solutions for the KdV Equation by Adomian Decomposition Method, Chaos, Solitons and Fractals, 12 (12): 2283-2293.

In: Nature Science and Sustainable Technology
Editor: M. R. Islam, pp. 137-166

ISBN: 978-1-60456-009-1
© 2008 Nova Science Publishers, Inc.

Chapter 7

A NOVEL SUSTAINABLE COMBINED HEATING/COOLING/REFRIGERATION SYSTEM

M. M. Khan[*,1], D. Prior[2] and M. R. Islam[1]

[1]Civil and Resource Engineering Department,
Dalhousie University, Halifax, Canada
[2]Veridity Env. Tech., Halifax, Nova Scotia, Canada

ABSTRACT

To date, the conventional calculation of coefficient of performance (COP) shows that the vapor compression refrigeration system is more efficient than absorption refrigeration system due to higher COP. However, a true pathway analysis shows that the conventional COP calculation is incomplete. This paper suggests a new way to calculate COP for comparing cooling systems. This new calculation shows that the refrigeration/cooling cycles for the same surroundings and cooling temperature levels, the COP of an absorption system is almost 2.5 times greater than that of a vapor compression. This result revives the interest of the use of absorption refrigeration system. Absorption refrigeration system is unique owing to its simplicity, compactness, and silent operation. However, the process is particularly appealing when it is coupled with direct usage of solar energy, which is the focus of the current study. Absorption cooling is already known as the heat dependent air conditioning system. On the other hand, both dual and triple pressure refrigeration cycles are not solely heat dependent. Only the single pressure refrigeration cycle is a thermally driven cycle that uses three fluids, as suggested in the Einstein refrigeration system. Calculations show that this system, coupled with the solar system, can be better than any other cooling system. The use of natural fluids and solar energy make the entire system inherently sustainable and economically attractive.

Key words: Coefficient of Performance, Solar cooling system, Sustainable technology

---

[*] Corresponding author: Email: mmkhan@dal.ca

INTRODUCTION

The objective of this paper is to design an energy-efficient refrigeration and air cooling system and compare it to the existing system. The ever-increasing demand of energy leads us to reduce the dependency of limited fossil fuel and to utilize alternative energy sources. Most existing processes are energy-inefficient, that is why much attention is needed to increase energy efficiency.

Two types of refrigeration system are commonly used today, namely, a) vapor compression type and b) absorption type. Vapor compression refrigeration cycle is an energy-inefficient technology, in which electricity is the source of energy. The energy consumption as well as energy loss of vapor compression cycle is very high. On the other hand, absorption type refrigeration/cooling system is operated by direct firing. Among different absorption refrigeration cycles, we are particularly interested on the single pressure thermally driven refrigeration cycle that is derived from a patent (US Patent: 1,781,541) issued to Einstein and Szilard. Most of the absorption cooling systems have at least one pump to lift the fluid but Einstein refrigeration is solely heat dependant. Delano (1998) in his research work created a thermodynamic model of the Einstein refrigeration cycle to calculate the cycle performance.

Around 1930, General Motors and Du Pont developed the first synthetic refrigerant (trade name: Freon) which was claimed nontoxic during that time (Schroeder, 1999). These chemical compounds are extremely stable, nonflammable, non-corrosive and cheap to produce and that is why they have been widely used as refrigerants especially in vapor compressor cooling systems (Auffhammer et al., 2005). However, scientists recognized as early as 1974 that the extensive use of Freon (CFC or HCFC) is depleting the Ozone Layer (Tsai, 2005). To reduce the destruction of stratospheric ozone a treaty known as The Montreal Protocol in 1987 was signed by 24 nations and the European Economic Community to regulate the production and trade of ozone-depleting substances and thereafter revised several times (Sozen el al., 2005; Tsai, 2005; and Auffhammer et al., 2005). The Montreal Protocol on Substances that Deplete the Ozone Layer (1987) is considered to be one of the most successful and important piece of international environmental legislation in history (Auffhammer et al., 2005).

It took nearly 60 years to ban Freon after realizing its detrimental effect to the earth. On the contrary, Albert Einstein and Leo Szilard patented the adsorption refrigeration system in 1930. Almost 65 years later, Delano (1998) first validated it and found the process sustainable. In fact, to-date, this process is one of very few sustainable processes, which uses environmentally benign fluids. It is one of the reasons to choose Einstein cooling system in addition to the use of direct solar heat.

Solar energy is the only ultimate source of energy available to the planet earth. This source is clean, abundant, and free of cost. The solar constant of solar energy is 1367.7 W/m^2 which is defined as the quantity of solar energy (W/m^2) at normal incidence outside the atmosphere (extraterrestrial) at the mean sun-earth distance (Mendoza, 2005). In space, solar radiation is practically constant; on earth it varies with the time of the day and year as well as with the latitude and weather. The maximum value on earth is between 0.8 and 1.0 kW/m^2 (Website 1). But the method of utilizing solar energy is different from one application to another. Even when solar energy is utilized, the mere fact that the most common usage is the use of photovoltaic, the maximum efficiency can be only 15-20%. The design of any system that operates under the direct use of solar energy has maximum energy conversion efficiency. The solar energy is an excellent source for thermally driven absorption refrigeration cycles. They do not require a costly electric power plant and they use environmentally benign natural fluids. Current absorption systems are dominated by dual-pressure cycles using a solution pump (which still requires a small electrical power source). Single-pressure cycles remove the need for a pump and any electrical power. Heat is the driving source for this heat pump.

The direct thermally driven refrigeration system offers the following advantages over vapor compression cycle (Cui et al., 2005):

1) Silent operation: Thermally driven absorption refrigeration does not have any moving parts (compressor) unlike vapor compression refrigeration system.
2) Simple structure: It can operate at fairly low pressure compare to vapor compression refrigeration cycle and that is why it does not need any high pressure equipment and piping system.
3) No need for electricity: Upon efficient design, absorption refrigeration system can solely be operated by heat. So no electricity is required.
4) Higher heating efficiency: Heat is directly used in the absorption system. So it eliminates the loss of heat during conversion heat to electricity. Due to direct use, higher heating efficiency is obtained.
5) Inexpensive equipment: No high pressure equipments, piping system and moving parts are required. Only some low pressure reservoir and piping system are required which are inexpensive compare to vapor cycle refrigerator.
6) No moving parts: As there is not compressor and moving parts, it is safer than other system.
7) High Reliability: Absorption system is highly reliable. It is not subjected to any electrical disturbance or any moving parts' problem.
8) Portability: The independence of electricity has made it portable. It can be installed in any place where there is a heat source. As solar energy is everywhere, it can be operated in any place.

These advantages make them ideal for remote locations as well as for places without electric utility infrastructures. They are also useful in any location where silent operation is essential. Direct use of solar energy as a heat source for the single pressure refrigeration cycle makes the refrigeration system completely energy efficient and independent of any other heat sources.

In this paper, the refrigeration system has taken as an example of how improvements can be made by taking advantage of the Einstein cycle and making use of direct solar irradiation. This refrigeration principle can be used both in a refrigerator and in an air cooler.

EINSTEIN REFRIGERATION CYCLE

All thermally driven heat pump cycles exchanging heat with only three temperature reservoirs are shown in Figure 1. Einstein cooling system is not an exception to this but the use of three fluids which circulate in different reservoir makes it solely heat dependent and unique. Einstein and Szilard proposed the use of butane, ammonia, and water as the working fluids in their suggested absorption refrigeration system (Einstein and Szilard, 1930).

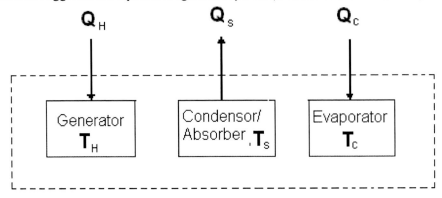

Figure 1. Thermally driven refrigeration system

This cycle is completely thermally driven that uses environmentally benign fluids (Alefeld, 1980; Dannen, 1997; and Delano, 1998). In the cycle butane acts as the refrigerant, ammonia as an inert gas, and water as an absorbent. It has three main components: a) evaporator b) condenser/absorber and c) generator. The fluid cycles in different reservoirs are depicted in the Figure 2.

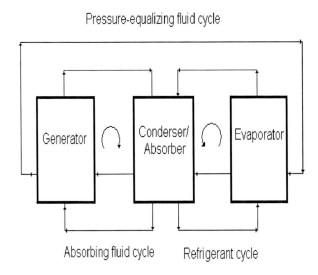

Figure 2. Fluid cycles in the Einstein refrigeration cycle

In the Einstein Cycle, Evaporator is fed with liquid butane from the absorber and gaseous ammonia from the generator (Fig. 3). The presence of ammonia in the evaporator reduces the partial pressure of butane that results in a decrease of saturation temperature. This causes the butane to evaporate and cools the system and surroundings. As soon as the gaseous mixture (butane and ammonia) reaches the condenser/absorber through pre-cooler, ammonia is readily absorbed by the spraying water coming from the generator. As a result, the partial pressure of butane increases nearly to the total pressure and it condenses in high saturation temperature at that total pressure. The immiscibility and lighter density causes butane to float on top of the liquid ammonia-water. Butane is siphoned back to the evaporator. The liquid ammonia-water leaves the absorber/condenser and reach to the generator through heat exchanger. Application of heat inside the generator drives off ammonia vapor to the evaporator. The remaining weak ammonia solution is pumped up to a reservoir via a thermally driven bubble pump. The ammonia from the reservoir is separated and sent to the absorber/condenser and the liquid from the reservoir, mainly, water is passed through heat exchanger and sprayed over superheated gaseous mixture and thus, the whole cycle is completed.

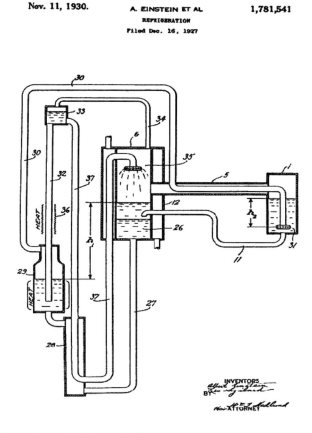

Figure 3. Einstein Refrigeration cycle (Delano, 1998)

THERMODYNAMIC MODEL AND THE ENERGY REQIRMENT OF THE CYCLE

Delano (1998) in his research work created a thermodynamic model of the Einstein refrigeration cycle to calculate the cycle performance. The cycle shown in Figure 4 has been modified from Einstein's original refrigeration cycle configuration. Various heat exchangers, such as the internal generator solution heat exchanger, evaporator pre-cooler have been added in order to create a cycle with a higher efficiency. After an extensive analysis, Delano (1998) fixed the system pressure at 4 bar to obtain realistic operating conditions. Delano (1998) used Patel-Teja equation of state to get the behavior of an ammonia-butane mixture at the evaporator and the behavior of ammonia-water mixture at generator.

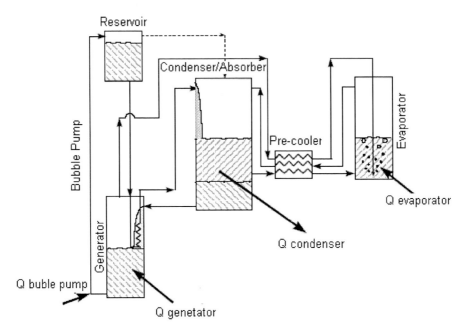

Figure 4. Einstein refrigeration cycle schematic (Redrawn from Shelton et al., 1999)

A temperature-concentration plot of ammonia-butane mixture obtained by Delano (1998) is shown in Figure 5. From this plot it is clear that for a given refrigerant (butane) at a fixed system pressure (4 bar), there is a maximum temperature (315 K) and a minimum temperature (266 K) to which the system can operate. These two extreme temperatures were chosen as the condenser/absorber temperature and as the evaporator temperature respectively for the thermodynamic model of Delano (1998).

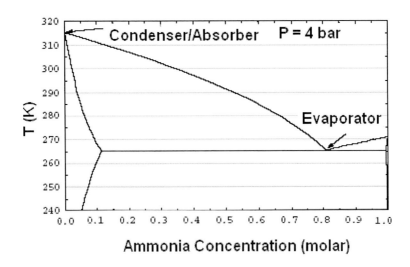

Figure 5. T-x-x-y diagram for ammonia-butane (Redrawn from Shelton et al. 1999)

Selection of a suitable generator temperature needs to establish a temperature-concentration diagram for ammonia-water mixture. A T-x-y diagram of ammonia-water mixture was obtained by Delano (1998), as shown in the Figure 6. Lower temperatures would reduce the amount of the desorbed ammonia vapor, and higher temperatures would boil unwanted water vapor (Fig. 6). After a detailed analysis, Delano (1998) selected 375 K as the maximum generator temperature to get a higher efficiency.

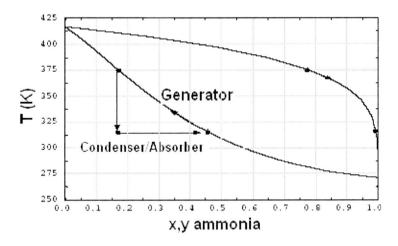

Figure 6. T-x-y diagram for ammonia-water, p =4 bar (Redrawn from Shelton et al. 1999)

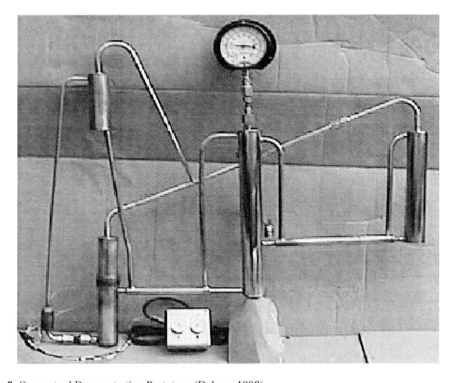

Figure 7. Conceptual Demonstration Prototype (Delano, 1998)

Energy and mass balance over each component provides sufficient information for the total cycle. The detailed calculation is given in the research works of Delano (1998). Delano

Energy and mass balance over each component provides sufficient information for the total cycle. The detailed calculation is given in the research works of Delano (1998). Delano (1998) also carried out a comprehensive first and second law of thermodynamics analysis of the cycle on each process to identify the thermodynamic sources of irreversibility and, therefore, the sources of the low efficiency. They identified generator and evaporator as the largest irreversibility.

The key component of this system is bubble pump that makes the whole system solely thermally driven without any electrical interference. Bubble pump is thermally driven pump that uses buoyancy lift arising from the thermal difference in the liquid reservoir relative to the liquid in the generator. The details of this pump have been discussed in the research works of Delano (1998). Delano (1998) demonstrated a prototype of Einstein cooling system and proved the practical viability of Einstein cooling system (Fig. 7).

The thermal energy needed to supply for Einstein refrigeration system can be supplied directly from solar energy and makes the Einstein refrigeration system more economical and environmentally friendly.

SOLAR COOLER AND HEAT ENGINE

Beside solar refrigerator, solar air cooler has received much attention recently. Most of the solar absorption air cooler systems found in the literature are single-stage two-fluid systems using either $LiBr/H_2O$ or H_2O/NH_3 as working fluids (Elsafty et al., 2002; Velazquez and Best, 2002; and Syed et al., 2005). The detailed analysis of those two-fluid systems shows that those systems require cooling water supply and a number of pumps. Those pumps create dependency on electricity if they are not solar pumps. Besides, cooling water supply needs an installation of cooling tower, which is recognized as a difficulty for maintenance (Li, and Sumathy, 2000).

Using same Einstein principle with some modification, the Einstein absorption refrigerator can be extended to be used as solar air cooler for space cooling at home. The additional parts of the air-cooling system compared to the refrigeration system are the chilled water storage and the chilled water circulation system that circulates chilled water from the evaporator to the cooling space. Air-condition system generally requires higher cooling capacity with 5 to 7° C cooling temperature. In the solar air cooler system, the condenser/absorber unit can be placed outside the room space, so that a better heat exchange is obtained due to temperature gradient and outside natural air circulation. Even though the evaporator temperature can be reached down to - 7° C, sufficient water circulation through the evaporator gives a chilled water storage temperature of 5 to 7° C. The water circulation in the room to be cooled can be facilitated with a solar pump. Extended surface to the water tube will enhance the heat absorption. Figure 8 shows an air conditioning unit. The same unit can be used as heat pump during winter if the absorption/condenser unit is kept in the room and evaporator is kept outside of the room. The absorption/condenser unit is considered as the heat transferring unit during winter.

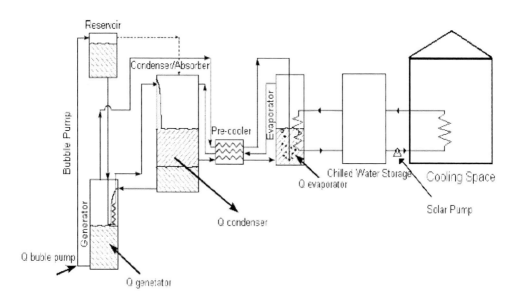

Figure 8. Air cooling system

ACTUAL COEFFICIENT OF PERFORMANCE (COP) CALCULATOIN

In this paper, the absorption refrigeration system and the vapor compression cycle refrigeration system have been compared to identify the true efficient system. To date, the efficient cooling system is identified by calculating its coefficient of performance (COP) (Smith et al., 2001).

The general definition for COP of a refrigerator can be defined as the ratio of the energy removed form the desired space to the energy required to drive the process.

For vapor compression cycle COP is defined as follows (Smith et al., 2001):

$$COP_v = \frac{\text{Heat (removed)}}{\text{Net work}} \qquad (1)$$

Whereas the COP of absorption air-conditioner is defined as the ratio of the heat transfer rate into the evaporator to the heat transfer rate into the generator (Li and Sumathy, 2000):

$$COP_a = \frac{\text{Heat transfer rate into the evaporator}}{\text{Heat transfer rate into the generator}} \qquad (2)$$

The energy balance from the primary energy source to the refrigeration system shows the definition of COP for vapor compression system does not indicate the true COP. It is found that in the vapor compression refrigeration system, the energy input is the work done by a compressor which is driven by electricity. Generally, this electricity comes from a power plant which is driven primarily by the heat. On the other hand, the COP calculation of absorption system includes heat as the primary source of energy. So, the vapor compression cycle is found as a process of converting heat into work and then using work to pump heat through a temperature lift, whereas the absorption system use heat to pump heat directly. The concern is that the absorption system elimination the requirement of a power plant and thus a better choice for the refrigeration system.

Vapor Compression Cycle Refrigeration System

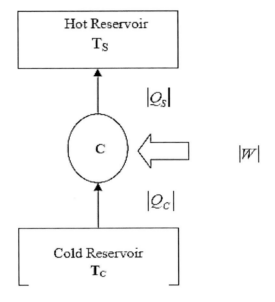

Figure 9. A schematic diagram Carnot refrigerator

Any vapor compression cycle that follows a Carnot cycle operating within two adiabatic steps and two isothermal steps in which heat $|Q_C|$ is absorbed at the lower temperature T_C requiring a net amount of work $|W|$, and heat $|Q_S|$ is rejected at the higher temperature T_S; the first law of thermodynamics therefore reduces to (Fig. 9):

$$|w| = |Q_S| - |Q_C| \qquad (3)$$

According to the Eq. (1) the COP can be written as:

$$COP_v = \frac{|Q_C|}{W} \qquad (4)$$

From the Carnot cycle it is found that (Smith et al., 2001),

$$\frac{|Q_S|}{|Q_C|} = \frac{|T_S|}{|T_C|} \qquad (5)$$

Using Equations (3), (4) and (5), COP can be further defined as:

$$COP_v = \frac{T_C}{T_S - T_C} \qquad (6)$$

This is the COP equation for the vapor compression refrigeration cycle system that follows the Carnot cycle.

Absorption Refrigeration System

Absorption refrigeration system operated in the three temperature levels (T_H, T_S, T_C) as shown in Figure 1. For the COP calculation, two pairs of temperature levels (T_H, T_S and T_S, T_C) are chosen in such a way as if two Carnot cycles are operating in those pairs.

The Carnot cycle that operated between temperature T_H (generator's temperature) and T_S (surrounding temperature) is a heat engine. The thermal efficiency of a Carnot heat engine that operated within two temperature levels (T_H and T_S) can be defined as (Smith et al., 2001):

$$\eta = \frac{\text{Net work ouput}}{\text{Heat absorbed}} \qquad (7)$$

$$= \frac{|W|}{|Q_H|} = \frac{|Q_H| - |Q_S|}{|Q_H|} \qquad (8)$$

$$= 1 - \frac{|Q_S|}{|Q_H|}$$

$$= 1 - \frac{T_S}{T_H} \qquad (9)$$

From Equations (8) and (9) the following equation can be written:

$$|Q_H| = |W| \frac{T_H}{T_H - T_S} \qquad (10)$$

Other Carnot cycle that is operated between temperature T_S (surrounding temperature) and T_C (evaporator's temperature) is a Carnot refrigerator.

From Equations (4) and (6) the following equation can be obtained:

$$W = \frac{T_S - T_C}{T_C}|Q_C| \tag{11}$$

Substituting W in the Eq. (11), the following equation can be obtained:

$$|Q_H| = |Q_C|\frac{T_H}{T_H - T_S}\frac{T_S - T_C}{T_C} \tag{12}$$

According to the definition of COP for absorption refrigeration:

$$COP_a = \frac{|Q_C|}{|Q_H|} = \frac{T_H - T_S}{T_H}\frac{T_C}{T_S - T_C} \tag{13}$$

From Equations (6), (9) and (13), the following relation can be found (Smith et al., 2001):

$$COP_a = \eta_{Carnot\,heat\,engine} \times COP_{Carnot\,vapor\,compression\,cycle} \tag{14}$$

It is found that for the absorption system the COP is equal to the multiplication of the efficiency of Carnot heat engine and the COP of Carnot vapor compression refrigeration system.

The COP calculation for the absorption system includes primary heating to finally cooling load. But the COP equation of the vapor compression cycle is not complete as it does not include the primary heating load. This exclusion makes the calculated COP of vapor cycle system higher than that of absorption system for the same cooling and surrounding temperatures. That is why conventional COP is not a true indication of cooling efficiency.

Identifying the above anomaly, this study has suggested a new way to calculate the COP for vapor cycle refrigeration system. According to this study, for an actual and complete COP calculation for the vapor cycle system, the global efficiency (η_{Global}) from primary heat input to the compressor output should be multiplied with the conventional COP so that the conversion of heat to electricity is counted. So, the propsed COP of the vapor compression system will be:

$$COP_{vp} = \eta_{Global} \times COP_v \tag{15}$$

The calculation of global efficiency is necessarey to obtain the theorytically true COP (proposed) of vapor compression system.

Calculation of Global Efficiency

A detailed analysis has been performed to calculate the efficiency from heating at boiler to the output of compressor. Figure 10 shows a widely used steam power plant which transforms heat energy to electrical energy.

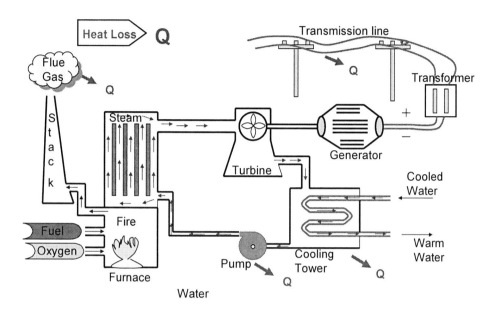

Figure 10. Typical steam power plant

This electricity is the input of compressor of the vapor compression cycle refrigeration system as shown in Figure 11.

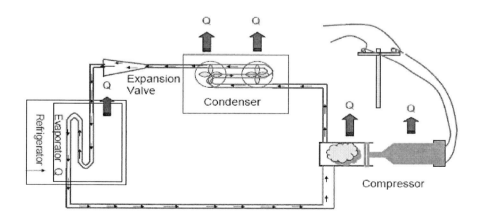

Figure 11. Vapor compression refrigeration system

To obtain the global efficiency of energy conversion, the efficiency of each component of the process should be calculated.

Heat Transfer Efficiency

Heat transfer efficiency indicates what fraction of heat is actually transferred to the vessel and water (Gupta et al., 1998). It measures the ability of the exchanger to transfer heat from the combustion process to the water or steam in the boiler. It includes the radiation and convection loss of the boiler and the loss of heat in the stack gas. Instead of heat transfer efficiency, it is better to use the term boiler efficiency or fuel-to-steam efficiency. Boiler efficiency is a measurement of how much combustion energy is converted into steam energy (Williams, 2003). With recent improvements, the boiler efficiency is found up to 70% (Barroso et al., 2003). The increase of access air up to a limit increases the combustion efficiency but it decreases the heat transfer efficiency as the access air takes more heat and looses through the stack gas.

Turbine Efficiency

A team turbine is a mechanical device that extracts thermal energy from pressurized steam, and converts it into useful mechanical work (Website 2). Turbine efficiency is depended on the thermal efficiency. The thermal efficiency of the engine is defined by equation (7) and (9). According to the second law of thermodynamics, Carnot's equation gives the theoretically maximum efficiency of any heat engine operation between two temperature levels.

The Carnot cycle comprises some systems and processes which have no existence in reality. A very simple analysis of the Carnot cycle shows the following impossibilities:

1) Ideal gas
2) Reversible process
3) Adiabatic process
4) Isothermal process (without phase change)

The definition of ideal gas shows that it is a hypothetical gas consisting of identical particles of negligible volume, with no intermolecular forces. So, there is no practical existence of an ideal gas. Similarly, No process can be perfectly reversible, adiabatic or isothermal, unless the process is extended infinitely in space – a far cry from other conventional 'closed' systems. So, the Carnot equation based on the above assumptions is merely a theoretical equation, which does not have any real basis.

Any heat engine that operates between 300° C and 30° C has a Carnot efficiency of 53%. But the actual efficiency (37%) is less than 70% of the maximum possible value when steam quality, friction loss and other non ideal conditions are considered. Kolev et al. (2001) reported that the overall efficiency of the turbine does not exceed 48.5% even for a very efficient system.

Generator Efficiency

The generator converts mechanical energy into electrical energy by the Faraday Effect. Most of the power plants produce AC current by means of alternators (synchronous generator). The generator efficiency of new large plants has been reported in excess of 90% (Weisman et al., 1985). The generator efficiency of average type of power plant can be taken as 80%.

Transmission Efficiency

After the generation of electricity, it is necessary to deliver it to the consumer. That is why a network is established to transmit and distribute electricity to the user's point. The network losses can represent 5%–10% of the total generation (Unsihuay et al., 2006). This number, however is grossly conservative. One significant loss of transmission is the heat loss due to the resistance of the transmission wire. The more the transmission line length from the origin, the more the energy loss.

Compressor Efficiency

For vapor cycle refrigeration of cooling system compressor is considered as the heart of that system. For reciprocating compressors, isentropic efficiencies are generally in the range of 70 to 90% (Peters and Timmerhaus, 1991). In this study, 80% is chosen as the practical compressor efficiency.

Global Efficiency

The global efficiency from primary heating to the cooling input includes the efficiency's of several units (Fig. 12).

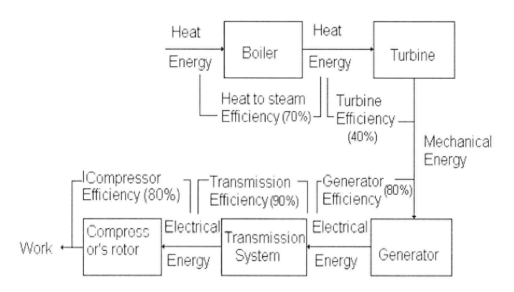

Figure 12. Efficiency calculation for a steam power plant up to cooling system

If we calculate the efficiency from heat transfer to the boiler water to the compressor's output of the cooling/refrigeration system, the global efficiency for a typical power plant will be:

Global heat transfer efficiency (η_{Global}) = Heat-to-Steam efficiency (70%) × Turbine efficiency or thermal efficiency (η_t) × Generator efficiency (80%) × Transmission efficiency (90%) × Compressor's rotor efficiency (80%) \hfill (16)

Global heat transfer efficiency (η_{Global}) = 40% × (η_t) \hfill (17)

From Equations (15) and (17), we get the new COP equation for the vapor compression cycle:

$$COP_{vp} = 40\% \times \eta_t \times COP_v \qquad (18)$$

On the other hand, if the turbine efficiency is taken η_t, the COP for the absorption system will be (from Eq. 14):

$$COP_a = \eta_t \times COP_v \qquad (19)$$

Comparing Eq. (18) and Eq. (19), it is found:

$$\frac{COP_a}{COP_{vp}} = \frac{\eta_t}{40\% \times \eta_t} = 2.5 \qquad (20)$$

So it is found that if a true pathway is analyzed and included, cycle for the same surroundings and cooling temperature levels the COP of an absorption system is almost 2.5 times greater than that of a vapor compression. This result indicates that direct use of heat has the maximum efficiency.

According to Eq. (13), any absorption refrigeration system, which operates with a generator temperature 100°C, condenser/absorption temperature 42°C and cooling temperature – 7°C, has a COP 0.84 which is theoretical maximum. Using equation (18) for the same operating conditions of the vapor compression refrigeration system, the proposed COP becomes 0.336. But the actual COP is found 30-40% of Carnot COP (Website 3). So, the actual COP of absorption refrigeration system and the vapor compression refrigeration system will be 0.336 and 0.134 respectively. The estimated COP of absorption refrigeration system is higher than that of vapor refrigreration system when true path is followed. According to this calculated value, for any absorption refrigeration systems having a cooling load 3.517 kW (1 ton), the heating load will be 10.47 kW (according to Eq.2). The choice of heating source is important.

Many sources of energy are present in the earth. Fossil fuels are considered as non-renewable energy source though they are renewable over very long period times. The increasing rate of fossil fuel utilization is alarming as fossil fuel is not readily renewed. So, only the dependency of the constant source of energy can eliminate the depletion of energy sources. Solar energy is considered as a constant source of energy. The energy extraction process from the source is different for different energy sources. The extraction efficiency of energy from its original source to heating is also an important factor to understand the efficiency of the extraction process.

Fossil Fuel

Combustion Efficiency

Energy is released from the fossil fuel by the combustion process. So, the efficiency of combustion is important to understand the efficiency of energy transfer. Combustion efficiency measures the extent to which the chemical energy of a fuel is converted in heat (Gupta et al., 1998). The amount of unburned fuel and excess air in the exhaust are used to assess the combustion efficiency of a furnace. A quality furnace design will allow firing at minimum excess air levels of 15% over the stoichiometric amount. Combustion efficiency varies from 50 to 90 % depending upon fuel specification, access air levels and furnace's design.

Solar Energy

For any solar high-temperature system, solar contractor is necessary. For the Einstein cooling system, parabolic solar trough is sufficient to meet the operating temperatures and heating requirement. This process travels only to few processes such as solar collector and the transmission.

Solar Collector Efficiency

The solar collector efficiency indicates the fraction of solar energy that can be transferred to the thermal fluid in the receiver. The parabolic solar collector efficiency varies much on the fluid temperature. Eck et al. (2005) reported that the collector efficiency shows higher at low temperature range (Fig. 13).

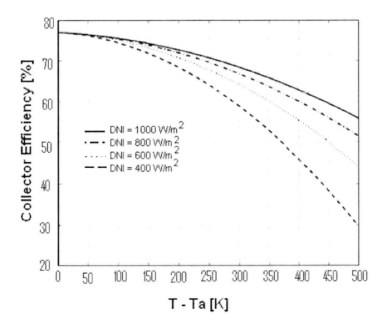

Figure 13. Collector efficiency at different direct normal irradiance (DNI) as a function of fluid temperature above the ambient temperature (Redrawn from Eck et al., 2005)

At low fluid temperature the thermal loss is minimum, as shown in Figure 14. From the figure, it is found that at a fluid temperature of 100°C (78°C above ambient temperature) the efficiency of solar collector is 75%.

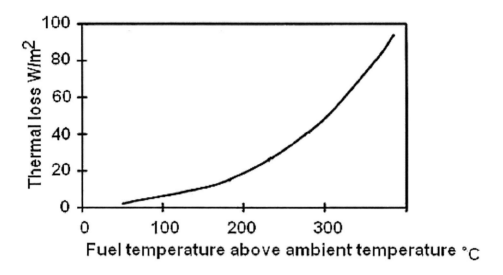

Figure 14. The thermal loss of the collector with respect to fluid temperature above the above the ambient temperature (Redrawn from Odeh et al., 1998)

The Einstein cooling system described in the paper has a maximum generator temperature nearly 100°C. So the solar collector can be operated at lower fluid temperature and thus increased the collector efficiency.

Transmission Efficiency
The solar transmission efficiency is dependent on the heat transfer loss from the thermal fluid to the fluid in the generator and the bubble pump. An efficient system will have more than 90% efficiency of transmission.

If the efficiency of solar system is calculated from the solar energy on the parabolic surface to the heat transfers to the refrigerator's generator fluid: the overall efficiency will be:

Overall energy transfer efficiency = Collector efficiency (75%) × Transmission efficiency (90%).

Overall energy transfer efficiency = 67.5 % (21)

It can be speculate that the extraction process of energy from different process does not differ much. So the consideration of solar system is beneficial as it has other benefits as discussed earlier.

SOLAR ENERGY UTILIZATION IN THE REFRIGERATION CYCLE

There are some existing efficient methods to concentrate the diluted solar energy and transfer to the desired places. The most common method is the use of a parabolic trough (Fig. 15) for the concentration of solar energy to obtain high temperature without any serious degradation in the collector's efficiency (Bakos et al., 2001; Geyer et al., 2002; and You et al., 2002). The solar refrigerator was first proposed by DeSa (1965) but that was not very successful one because of poor solar collector. Later on, much improvement on the solar collector has been noticed. The reflector, which concentrates the sunlight to a focal line or focal point, has a parabolic shape. The parabolic trough collector consists of large curved mirror, which can concentrate the sunlight by a factor of 80 or more to a focal line depending upon the surface area of the trough. In the focal line of these is a metal absorber tube, which is usually embedded into an evacuated glass tube that reduces heat losses (Fig. 16). A special high-temperature, resistive selective coating additionally reduces radiation heat losses.

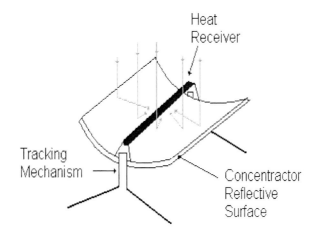

Figure 15. Parabolic Trough

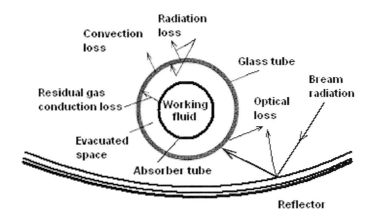

Figure 16. Cross section of collector assembly (Redrawn from Odeh et al., 1998)

California power plants, known as solar electric generating system has a total installed capacity of 354 MW (Kalogirou et al., 1997). This system uses thermo-oil as a heat transfer fluid, which can reach up to 400°C (Herrmann et al., 2004). The parabolic collector effectively produces heat at a temperature between 50°C and 400°C (Kalogirou, 2004).

THE NEW SYSTEM

In this study, a parabolic trough has been constructed that is adjustable and moves along the direction of sun so that maximum solar energy can be achieved anytime of the day (Fig. 17). Each parabolic trough has a surface area of 4 m^2 (2.25 m × 1.8 m) that can radiate almost 1.6 kW to 4 kW to the absorber depending on the direct normal irradiance, which is again dependent on the geographical area. Taking 600 w/m^2 as DNI (direct normal irradiance) and considering the energy transfer efficiency (Eq.21) from solar surface to the heating point, it is found that one surface (4 m^2) can supply 1.62 kW. So, a heating load of 10.47 kW will require 7 such parabolic collectors, which can supply necessary energy to run a refrigerator or an air cooler having one ton cooling load. The number of collectors will vary from place to place depending on the DNI of any place and the climate of that place. The experimental data shows that the parabolic collector can absorb 0.80 kW during early summer in a cold country when the environmental temperature is nearly 21°C. This parabolic trough can be placed in the roof of house or can be mounted in the outside wall of house where the availability of sunlight is the highest. A proper insulation of the carrier tube can reduce the heat loss of the transferring process. The thermal fuel (vegetable oil) is circulated by the solar pump and that is why no electricity is needed (Fig. 18).

A household kitchen based biogas production can enhance the operability of this refrigeration in the absence of sun. Biogas can be burnt to get the necessary heat to operate the cooling system especially at night.

In this study, toxic thermal oil has been replaced to environmentally friendly vegetable oil. This study claims the first reporting of using vegetable oil as thermal oil.

Figure 17. Constructed parabolic trough

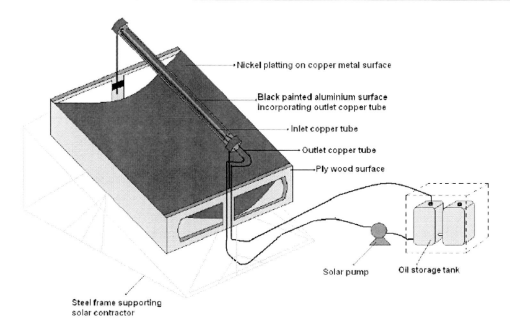

Figure 18. Experimental solar trough (Redrawn from Khan and Islam, 2006a)

PATHWAY ANALYSIS

Most existing processes are energy-inefficient that is why much attention is needed to increase energy efficiency. To obtain the true efficiency, it is needed to analysis the pathway of any process. Recently, Islam et al. (2006) identified three basic factors of energy which need to be analyzed before addressing a process as an efficient process. These three factors are:

a) Global economics
b) Environmental impact
c) Quality

Starting from the extraction of energy from the source to its application and its affect to the products and the consumers should be carefully analyzed to identify the efficient process.

Environmental Pollution Observation

Energy and environment are two of the most concerning issues in the current world. Today most electricity is generated by consuming fossil fuels such as coal, oil and natural gas. Not only do fossil fuels have a limited life but also their combustion emissions have serious negative impacts on our environment such as adding to the greenhouse effect and causing acid rain. That is why, for any process, the pollution should be considered as an important factor which is eluded in most of the cases. For some process, only the consideration of pollution can be the decisive factor. For the steam power plant process, the pollution and loss of heat can be analyzed by considering the following stages.

- Fuel processing stage
- Combustion stage
- Transmission stage
- Refrigeration stage

Fuel Collection Stage

Toxic drilling fluids, synthetic drilling mud, toxic surfactant addition or chemical treatment for EOR project etc cause great environmental pollution during the oil production from the reservoir. This oil needs to be refined before using in the steam power plant. During this refining process lots of toxic materisl are added, which make the fuel environmentally vulnerable. Chhetri et al. (2006a) described the toxicity addition of a fuel during processing stage (Fig. 19)

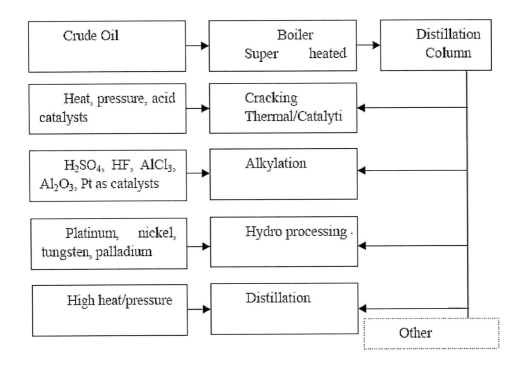

Figure 19. Pathway of oil refining process (Redrawn from Chhetri et al., 2006a)

Besides, produce oil and gas from the reservoir reduces the necessary reservoir pressure, leading to significant subsidence. Recent investigation shows that this subsidence is not negligible.

Combustion Stage

In the absence of access air, the fuel can be burnt incompletely and produced carbon-monoxide. All the toxic gases realize through the stack gas greatly pollute the environment.

The vulnerable effect increases with the treated fuel. The more the treatment of the fuel, the more it is likely to cause pollution during combustion stage. At this stage, the toxic constituents can escape from the fuel and mix with the air through the stack gas. Chhetri et al. (2006a) have reported that burning of untreated fuelwood cause less damage to the environment that the treated fuelwood. The addition of different toxic material during fuel processing unit increases this tendency.

Transmission Stage

It is now a concern whether human exposure to power-frequency electromagnetic fields has significant health consequences (Morgan et al., 1985). Hamza et al. (2002) speculate that a source of environmental pollution is the magnetic field produced near high and extra high voltage (EHV) transmission lines. It has become a controversial issue as there no detailed research work is found to address this problem. Qualitatively, some researchers found it as a source of causing certain types of cancers (Website 4).

Refrigeration Stage

Vapor compression cooling system generally uses synthetic refrigerant fluids. The negative impacts of chlorofluorocarbons (CFC) and hydrochlorofluorocarbons (HCFC) are well documented. Besides participating in the destruction of stratospheric ozone, the release of CFCs may also contribute to global warming, which means that CFCs influence the reflection of infrared radiation from the surface of the earth and thus cause global climate change (Hayman et al., 1997). Because of similar physiochemical properties, HCFCs have been used as the replacement of CFCs. The release of chlorine from the HCFCs to the stratosphere has been identified detrimental to the earth for long term. According to the updated Montreal Protocol, a virtual phase-out of HCFCs is scheduled by 2020 (Tsai, 2005).

Research on the replacement of one synthetic refrigerant by another synthetic refrigerant is continuing. Replacement is taking place when it is assumed safe to the environment. But previous experience indicated that no synthetic refrigerants were found environmentally safe ultimately. It was claimed safe during replacement, but ultimately, it was found detrimental to the environment. Sometimes it is delayed to understand the vulnerable effect due to lack of appropriate knowledge. That is why it is safe to use natural refrigerant fluid instead of any synthetic refrigerant.

Thus, if all these pollution stages are considered, it can be found that the vapor compression cycle is source of environmental pollution. The pollution can be in different phases depending upon the fuel system and the refrigerant fluids. Radiation, greenhouse pollution, air/water pollution, disruption of ecosystem and the sickness of the human being are the direct and indirect effects of this system.

Further more, there are heat losses in every steps of this process, which suffers from a huge loss of heat. This is truly an indication of generating more heat for removing heat as shown in Figure 12 and Figure 14.

Environmentally Friendly System

Solar absorption Einstein cooling system is found to be an environmentally friendly system. The solar collection system doesn't produce any negative impact to the environment because solar energy is clean. Again the use of vegetable or waste vegetable oil makes it more environmentally acceptable.

Due to ozone-depletion and global-warming effects, environmentally friendly refrigerants with zero ozone-depletion potential are required to be used in refrigerators and heat pumps (Sozen el al., 2005; Fernandez et al., 2001; Hwang et al., 2002; and Zhang et al., 2002). The cooling system is found to use naturally benign fluids which don't have any significant long term effect upon environment.

Global Economics of the Systems

It is important to find a mean to calculate the global economics of any process. It has already been shown that the vapor compressor cooling systems involve in the pollution of the environment in different directions. Again, this is the process of energy depletion of the necessary concentrated energy sources. If the costs of the remedial process of all the vulnerable effects along with the plant cost are considered, the actual cost of the total process can be obtained. The cost involves the remedial cost of soil, the remedial cost of air/water pollution, the remedial cost of ecological lost, the cost of medical and medicinal for the human being etc (Zatzman and Islam, 2006). There are some more costs which are difficult to understand as the direct effect of the process involved.

On the other hand, the pathway of absorption cooling system indicates that it is not associated with the above vulnerable effects and that is why no additional cost is required.

Quality of Energy

The quality of energy bears an important phenomenon which is little understood. For example, heating of home by wood is better than heating by electricity. The radiation due to the electro-magnetic rays might interference with the human's radiation frequency which can cause acute long term damage to the human beings. Energy with natural frequency is the most desirable. Alternate current is not natural and that's why there will be some vulnerable effects of this frequency to the environment and the human beings (Chhetri et al., 2006b).

Even the cooling by using synthetic fluid, the leaky of this fluid might destroy the quality of the food subject to cooling.

Considering all the effect, the vapor refrigeration system seems an inefficient system, compare to solar absorption system.

SUSTAINABILITY ANALYSIS

Most of the technologies in use today are not beneficial to living beings. Even after extensive development of different technologies from decade to decades, it is found that the world is becoming a container of toxic materials and loosing its healthy atmosphere continuously. That is why, it is necessary to test the sustainability of any process and the pathway of the process. In this study, we have used the sustainability test model proposed by Khan and Islam (2006) to analyze the two different systems of cooling. According to this model, if and only if, a process travels a path that is beneficial for an infinite span of time, it is sustainable; otherwise the process must fall in a direction that is not beneficial in the long run. The pro-nature technology is the long-term solution, while anti-nature one is the result of Δt approaching 0. The most commonly used theme is to select technologies that are good for t='right now', or $\Delta t = 0$. In reality, such models are non-existent and, thus, aphenomenal and cannot be placed on the graph (Fig.20). However, "good" technologies can be developed following the principles of nature. In nature, all functions or techniques are inherently sustainable, efficient and functional for an unlimited time period. In other words, as far as natural processes are concerned, *'time tends to Infinity'*. This can be expressed as t or, for that matter, $\Delta t \to \infty$. For example, Freons were extremely stable, nonflammable, non-corrosive and cheap to produce at the beginning but its detrimental effects were understood after a long time and thus proved as unsustainable products.

Mathematically, the expression can be written as:

$$B = b + (\pm a)^i t e^{kt} \tag{22}$$

Where B indicates the benefit of the process with time. b is the initial condition of the process or the product to be analyzed. The value of b can be positive, negative or zero depending on the initial perception of the process or product. Symbols a and k are constants depending on various factors such as social, environmental and ecological factors of the products or processes to be analyzed. The index 'i' is considered to be the directional parameter, which depends only on intention and indeed certify whether the technology will be sustainable or not.

The positive and negative values indicate sustainable and unsustainable technology. The slope of the above equation can be written as:

$$\frac{dB}{dt} = (\pm a)^i k e^{kt} \tag{23}$$

The positive slope indicates the rate at which benefit is achieved. Similarly, a negative slope demonstrates how fast a process/technology will lead to un-sustainability. It must be added here that if intention is intended for long-term, even meant for self-interest, can be described by the positive slope. However, any action serving self-interest for the short term is bound to collapse and must be described by the negative slope in the figure. For any new technology to be tested mostly depends on the value of 'i' but for any existing technology that is already in operation can be tested from the field data.

Figure 20 has the inherent assumption that nature is perfect. Even though this concept is as old as the universe, only recently the science of this concept is being discussed (TIME, 2005). Because, no mass can be created or destroyed and no mass can be isolated from the rest of the universe; any entity must play a role in creating overall balance (Khan et al., 2006). This is where intention becomes important for human beings. An intention to benefit others at present time translates into multifold benefit for the subject of the intention. Therefore, benefit for infinite number of entities in space at a given time is equivalent to multifold benefit for the individual in the long run. The mathematics and physics of this argument are discussed elsewhere (Khan, 2006; Khan and Islam, 2005a and 2005b; Khan and Islam, 2006a; 2006b).

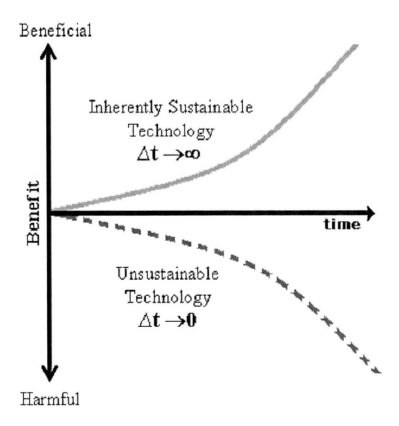

Figure 20. Direction of sustainability (Redrawn from Khan and Islam, 2006a)

The proposed model of this study has been tested for sustainability according to the above definition and details analysis has been presented here.

The technology that exists in nature must be sustainable, because natural process is already proved to be beneficial for long term. Ammonia, butane and water exist in nature. So the uses of these materials make the refrigeration process sustainable. Solar energy is non-toxic and free source. The utilization of solar energy must be a sustainable thinking. In this experimental solar collector, it has been recommended to use waste vegetable oil. The utilization of vegetable oil keeps the whole equipment corrosion free. Even any use of solar energy in turns reduces the dependency of fossil fuel and thereby saving the world from more air pollution. So this is also sustainable technology.

On the other hand, alternate electricity is not natural. The process of producing electricity poses vulnerable effect in every stage. The use of synthetic refrigeration is also ant-nature. These entire things make the vapor refrigeration system aphenomenal.

CONCLUSION

A complete analysis of the pathway is only the true identification of the efficiency of any process. The refrigeration cycle that operates completely upon thermal energy is the best choice for the energy efficient refrigeration system. Einstein cycle offers one such choice. However, there are many other types of absorption refrigeration cycles where solar energy can be utilized efficiently. Some changes in the design of energy uptake will be enough to make any single pressure refrigeration cycle solely solar energy dependent. The utilization of refrigeration system has been increased dramatically. Most of the refrigerators are built on vapor compression cycle driven by electricity. That is an unacceptable waste of energy. The pathway analysis of the vapor compression cooling system showed that the process is also unacceptable for healthy environment and true economical value. On the other hand, the solar energy driven refrigeration cycle is highly energy conservative and independent of fossil fuel. The whole system is environment-friendly and energy efficient. The maintenance cost being practically nil, the system is economically attractive.

ACKNOWLEDGEMENT

The authors would like to acknowledge the funding from the Atlantic Innovation Fund (AIF).

REFERENCES

Alefeld, G., 1980. Einstein as Inventor, *Physics Today,* May, pp 9-13

Auffhammer, M., Morzuch, B.J. and Stranlund, J.K., 2005. Production of Chlorofluorocarbons in Anticipation of the Montreal, *Protocol Environmental and Resource Economics*, vol. 30, no. 4, April, pp 377-391

Bakos, G.C., Ioannidis, I., Tsagas, N.F. and Seftelis, I., 2001. Design, Optimization and Conversion-Efficiency Determination of a Line-Focus Parabolic-Trough Solar-Collector (PTC), *Applied Energy,* vol. 68, no. 1, January, pp 43-50

Barroso, J., Barreras, F., Amaveda, H. and Lozano, A., 2003. On the Optimization of Boiler Efficiency Using Bagasse as Fuel, *Fuel,* vol. 82, no. 12, August, pp 1451-1463

Chhetri, A.B., Khan, M.I. and Islam, M.R., 2006a. A Novel Sustainably Developed Cooking Stove, *Journal of Nature Science and Sustainable Technology*: Submitted

Chhetri, A.B., Zaman, M.S and Islam, M.R., 2006b. Characterization of Energy Sources Based on Their Values, *Journal of Nature Science and Sustainable Technology*: Submitted.

Cui, Q., Tao, G., Chen, H., Guo, X. and Yao, H., 2005. Environmentally Benign Working Pairs for Adsorption Refrigeration, *Energy*, vol. 30, no. 2-4, SPEC. ISS., February/March, pp 261-271

Dannen, G., 1997. The Einstein-Szilard Refrigerators, *Scientific American*, January, pp 90-95

Delano, A., 1998. Design Analysis of the Einstein Refrigeration Cycle, Ph.D. Thesis, Georgia Institute of Technology, Atlanta, Georgia

DeSa, V.G., 1964. Experiments with Solar-Energy Utilization at Dacca, *Solar Energy*, vol. 8, no. 3, July-September, pp 83-90

Eck, M. and Steinmann, W.D., 2005. Modelling and Design of Direct Solar Steam Generating Collector Fields, *Journal of Solar Energy Engineering, Transactions of the ASME*, vol. 127, no. 3, August, pp 371-380

Einstein, A. and Szilard, L., 1930. US Patent No. 1,781, 541.

Elsafty, A. and Al-Daini, A.J., 2002. Economical Comparison Between a Solar Powered Vapour Absorption Air-Conditioning System and a Vapour Compression System in the Middle East, *Renewable Energy*, vol. 25, pp 569–583

Fernandez, S. and Vazquez, M., 2001. Study and Control of the Optimal Generation Temperature in NH_3–H_2O Absorption Refrigeration Systems, *Applied Thermal Engineering*, vol. 21, pp 343–357

Geyer, M., Lüpfert, E., Osuna, R., Esteban, A., Schiel, W., Schweitzer, A., Zarza, E., Nava, PLangenkamp, J. and Mandelberg, E., 2002. Euro Trough - Parabolic Trough Collector Developed for Cost Efficient Solar Power Generation, presented at: *11th Int. Symposium on Concentrating Solar Power and Chemical Energy Technologies*, Zurich, Switzerland, September 4-6

Gupta, S., Saksena, S., Shankar, V.R. and Joshi, V., 1998. Emission Factors and Thermal Efficiencies of Cooking Biofuels from Five Countries, *Biomass and Bioenergy*, vol. 14, no. 5-6, pp 547-559

Hamza, A.S. H.A., Mohmoud, S. A. and Ghania, S. M., 2002. Environmental Pollution by Magnetic Field Associated with Power Transmission Lines, *Energy Conversion and Management*, vol. 43, no. 17, November, pp 2443-2452

Hayman, G. and Derwent, R.D., 1997. Atmospheric Chemical Reactivity and Ozone-Forming Potentials of Potential CFC Replacements, *Environment Science Technology*, vol. 31, pp 327

Herrmann, U., Kelly, B. and Price, H., 2004. Two-Tank Molten Salt Storage for Parabolic Trough Solar Power Plants, *Energy*, vol.29, pp 883–893

Hwang, Y. and Radermacher, R., 2002. Opportunities with Alternative Refrigerants, *Thermomechanical Phenomena in Electronic Systems -Proceedings of the Intersociety Conference*, pp 777-784

Islam, M.R., 2006, Unraveling the Mysteries of Chaos and Change: Knowledge-Based Technology Development", *EEC Innovation*, vol. 2, no. 2-3, pp 45-87

Islam, M.R, Zatzman, G.M. and Shapiro, R., 2006. Energy Graph: What Mere Lies Ahead? *Energy Dialogue*, CSIS, Washington DC., April 4-5

Kalogirou, S., Lloyd, S. and Ward, J., 1997. Modelling, Optimisation and Performance Evaluation of a Parabolic Trough Solar Collector Steam Generation System, *Solar Energy*, vol. 60, no. 1, pp 49-59

Kalogirou, S. A., 2004. Solar Thermal Collectors and Applications, *Progress in Energy and Combustion Science*, vol. 30, no. 3, pp 231-295

Khan, M.I., 2006. Towards Sustainability in Offshore Oil and Gas Operations, Ph.D. Dissertation, Dalhousie University, Nova Scotia, Canada

Khan, M.I. and Islam, M.R., 2006a. True Sustainability in Technological Development and Natural Resources Management, Nova Science Publishers, New York, USA: in press.

Khan, M.I. and Islam, M.R., 2006b. Handbook of Sustainable Oil and Gas Operations Management, Gulf Publishing Company, USA: in press.

Khan, M.I., Chhetri, A.B. and Islam, M.R., 2006. Achieving True Technological Sustainability: Pathway Analysis of a Sustainable and an Unsustainable Product. *Nat.Sci. and Sust.Tech.* Vol. 1, no. 1.: in press..

Khan, M.I, Zatzman, G. and Islam, M.R., 2005. New Sustainability Criterion: Development of Single Sustainability Criterion as Applied in Developing Technologies. *Jordan International Chemical Engineering Conference V*, Paper No.: JICEC05-BMC-3-12, Amman, Jordan, September 12-14

Khan, M.I., Chhetri, A.B. and Islam, M.R., 2006a. Community-Based energy Model: A Novel Approach in Developing Sustainable Energy. *Energy Sources: in press.*

Khan, M.M, Zatzman, G. and Islam, M.R, 2006, The Formulation of Comprehensive Mass and Energy Balance Equation: Towards Modeling Negative Entropy. *Journal of Nature Science and Sustainable Technology,* Submitted.

Kolev, N., Schaber, K. and Kolev, D., 2001. New Type of a Gas-Steam Turbine Cycle with Increased Efficiency, *Applied Thermal Engineering*, vol. 21, no. 4, March, pp 391-405

Li, Z.F. and Sumathy, K., 2000. Technology Development in the Solar Absorption Air-Conditioning Systems, *Renewable and Sustainable Energy Reviews*, vol. 4, pp 267-293

Mendoza, B., 2005. Total Solar Irradiance and Climate, *Advances in Space Research*, vol. 35, pp 882–890

Morgan, M. G., Florig, H. K., Indira N. and Lincoln, D., 1985. Power-Line Fields and Human Health, *IEEE Spectrum*, vol. 22, no. 2, February, pp 62-68

Odeh, S.D., Morrison, G.L. and Behnia, M., 1998. Modelling of Parabolic Trough Direct Steam Generation Solar Collectors, *Solar Energy,* vol. 62, no. 6, June, pp 395-406

Peters, M.S. and Timmerhaus, K.D., 1991. Plant Design and Economics for Chemical Engineers, Fourth Edition, McGraw-Hill, New York, USA

Schroeder, D.V., 1999. An Introduction to Thermal Physics, Addison Wesley Longman, pp 137

Shelton, S. V., Delano, A. and Schaefer, L. A., 1999. Second Law Study of the Einstein Refrigeration Cycle, *Proceedings of the Renewable and Advanced Energy Systems for the 21st Century*, Lahaina, Maui, Hawaii, April 11-15

Smith, J.M., Van Ness, H.C. and Abbott, M.M., 2001. Introduction to Chemical Engineering Thermodynamics, Sixth Edition, McGraw-Hill, New York, USA

Sozen, A. and Ozalp, M., 2005. Solar-Driven Ejector-Absorption Cooling System, *Applied Energy,* vol. 80, no. 1, January, pp 97-113

Syed, A., Izquierdo, M., Rodriguez, P., Maidment, G., Missenden, J., Lecuona, A. and Tozer, R., 2005. A Novel Experimental Investigation of a Solar Cooling System in Madrid, *International Journal of Refrigeration,* vol. 28, pp 859–871

Tsai, W., 2005. Environmental Risk Assessment of Hydrofluoroethers (HFEs), *Journal of Hazardous Materials,* vol. 119, no. 1-3, March, pp 69-78

TIME, 2005, The Science of Happiness, Feb. 7, pp 38-59

Unsihuay, C. and Saavedra, O, R., 2006. Transmission Loss Unbundling and Allocation Under Pool Electricity Markets, *IEEE Transactions on Power Systems,* vol. 21, no. 1, February, 2006, pp 77-84

Velazquez, N. and Best, R., 2002. Methodology for the Energy Analysis of an Air Cooled GAX Absorption Heat Pump Operated by Natural Gas and Solar Energy, *Applied Thermal Engineering,* vol. 22, pp 1089–1103

Website 1., http://www.solarserver.de/lexikon/solarkonstante-e.html, accessed on 22th June, 2006

Website 2., http://en.wikipedia.org/wiki/Steam_turbine, accessed on 22th June, 2006

Website 3., http://www.electronics-cooling.com/html/2001_august_a3.html, accessed on 22th June, 2006

Website 4., http://en.wikipedia.org/wiki/Electric_power_transmission, accessed on 22th June, 2006

Weisman, J. and Eckart, L. E., 1985. Modern Power Plant Engineering, Pretice-Hall, Inc., NJ, USA, pp 410

Williams, C., 2003. Boiler Efficiency vs. Steam Quality, *HPAC Heating, Piping, Air Conditioning Engineering,* vol. 75, no. 1 SUPPL., January, pp 40-45

You, Y. and Hu, E. J., 2002. A Medium-Temperature Solar Thermal Power System and Its Efficiency Optimization, *Applied Thermal Engineering,* vol. 22, no. 4, Mar, pp 357-364

Zatzman, G.M. and Islam, M.R., 2006. Economics of Intangible, Nova Science Publishers, New York, *in press.*

Zhang, X. J. and Wang, R.Z., 2002. A New Adsorption–Ejector Refrigeration and Heating Hybrid System Powered by Solar Energy, *Applied Thermal Engineering,* vol. 22, pp 1245–1258

INDEX

A

abatement, 74
academics, 7, 8
accelerator, 84
access, 1, 81, 86, 150, 153, 158
accidents, 13
accounting, 15
accuracy, 39, 67, 119, 120, 124, 125, 126, 132, 134, 135
acetylene, 96
acid, 78, 79, 80, 96, 104, 105, 157
acidic, 108
adamantane, 75
additives, 89, 93, 94, 105, 108, 112
adiabatic, 91, 101, 146, 150
administration, 101
adsorption, 2, 73, 74, 76, 138
aerosols, 96, 97
age, 4, 35, 84, 85, 94, 98
agent, 95, 99
aging, 20, 94
agriculture, 91, 100
AIDS, 13
air pollution, 163
alcohol(s), 1, 13, 78, 107
algae, 82
alkaline, 37, 74
alternative, 4, 19, 35, 104, 138
alternative energy, 138
aluminum, 89, 92
Alzheimer's disease, 20
amide, 80, 81
amine(s), 78, 80
amino, 74, 75, 78, 79
amino acid(s), 78, 79
ammonia, 4, 20, 89, 96, 102, 140, 142, 143
ammonium, 75
analytical techniques, 40
animals, 22
anisotropy, 80
anthropogenic, 83, 84, 88, 113
anti-biotics, 1
antioxidant(s), 93, 94, 103
appetite, 87
argument, 3, 53, 55, 56, 57, 84, 85, 91, 162
arsenic, 76, 94, 99, 108
arthritis, 20
ash, 75, 94, 98, 99, 105, 108, 109
Aspartame, 95
assessment, 6, 10, 19, 23, 30, 35, 37, 88
assessment models, 10
assumptions, 40, 53, 54, 83, 91, 113, 150
asthma, 1, 20
atmosphere, 83, 84, 85, 86, 87, 88, 89, 90, 91, 92, 94, 96, 97, 101, 102, 107, 111, 138, 161
attention, 138, 144, 157
Australia, 100
availability, 85, 156

B

bacterium (a), 4, 94
batteries, 112
battery, 113
behavior, 19, 20, 30, 70, 73, 96, 142
benefits, 6, 20, 101, 105, 154
benign, 87, 108, 138, 139, 140, 160
benzene, 89, 105, 112
Bessel, 127
beverages, 14
bias, 17, 87
binding, 74
biodegradable, 6
biodiesel, 99, 104, 105, 112

biodiversity, 38, 107, 113
bioethanol, 105, 106, 112
biofuel(s), 20, 112
biomass, 1, 74, 82, 89, 90, 92, 105, 106, 111, 112, 113, 117
biopolymer(s), 74, 80, 82
bioremediation, 36
biosorption, 73, 82
biosphere, 8, 37
blocks, 43, 50, 64
bonding, 74, 80
boundary conditions, 39, 40, 42, 58, 59, 63, 67, 119, 120, 121, 128
brain, 3, 16, 20
Brazil, 8
breakdown, 105
breast milk, 6, 36, 37, 38
breeding, 13
Britain, 14, 86
brominated flame retardants, 34
bromine, 92
burn, 111, 117
burning, 88, 90, 91, 92, 94, 98, 99, 101, 105, 108, 109, 111, 113, 159
business model, 33
by-products, 97, 98, 107, 111, 113

C

cadmium, 82
calcium, 75, 80, 81, 98, 99, 103
calcium carbonate, 80, 81
California, 37, 156
calorie, 8
Canada, 5, 13, 35, 70, 73, 74, 82, 83, 87, 89, 91, 100, 103, 114, 116, 118, 119, 135, 137, 165
cancer, 1, 20, 37, 94, 103
cancer cells, 94
capillary, 70
carbohydrate, 1
carbon, 73, 75, 79, 80, 83, 84, 85, 87, 89, 91, 93, 94, 99, 101, 111, 113, 115, 117, 158
carbon dioxide, 80, 83, 84, 91, 93, 94, 101, 111
carbon tetrachloride, 93
carbonyl groups, 80
carboxyl, 80
carboxylic acids, 78
carcinogen(s), 6, 92, 103
Caribbean, 16
Carnot, 146, 147, 148, 150, 152
carotene, 94
carrier, 156
cartels, 15, 87

case study, 37, 117
casting, 92
catalysis, 87
catalyst(s), 85, 91, 98, 99, 104, 105, 106, 108, 111, 112
cation, 80, 82
cell, 82
cement, 89, 98, 108
CERES, 9
certification, 101
chaos, 35
chaotic, 18, 19, 20
chaotic behavior, 19
chemical, 70, 75, 78, 80, 82, 87, 89, 93, 94, 95, 96, 98, 99, 103, 107, 113, 138, 153, 158
chemical composition, 96
chemical energy, 153
chemical engineering, 87
China, 8, 101
Chinese, 8, 115
chitin, 74, 75, 80, 82
chitosan, 74, 75, 82
chlorine, 92, 95, 159
chloroform, 93
chlorophenols, 82
cigarette smoke, 13
circulation, 15, 144
classes, 92
classical economics, 15
classification, 100
clay, 111
Clean Development Mechanism, 83, 100, 113
clean energy, 101
cleaning, 95, 108, 109
climate change, 83, 84, 85, 86, 94, 96, 97, 100, 102, 113, 159
CO_2, 22, 83, 84, 85, 87, 88, 89, 90, 91, 92, 93, 94, 95, 96, 98, 99, 100, 101, 102, 105, 106, 107, 108, 109, 111, 112, 113, 114, 115, 118
coal, 14, 88, 89, 108, 112, 113, 157
coefficient of performance (COP), 137
coke, 89
collaboration, 31, 32
collagen, 78, 82
colonization, 86
combustion, 35, 89, 91, 94, 96, 98, 99, 106, 107, 108, 111, 112, 113, 150, 153, 157, 159
commercial, 13, 16, 66, 86, 91, 95
commodity(ities), 15, 87
communication, 28, 34
community, 29, 31, 36, 37
competition, 86
complement, 66, 67

components, 4, 9, 12, 31, 33, 73, 75, 94, 96, 97, 112, 123, 140
composite, 73, 80
composition, 75, 96
compounds, 6, 13, 30, 31, 87, 92, 103, 108, 113, 138
computation, 124, 132, 133
concentrates, 155
concentration, 14, 88, 89, 90, 92, 93, 99, 101, 102, 142, 143, 155
condensation, 96
conditioning, 8, 137, 144
conduction, 100
conductor, 92
configuration, 141
confusion, 11
Congress, 35, 36, 37, 116
conservation, 28, 42
consumer goods, 15
consumers, 157
consumption, 7, 12, 13, 32, 84, 87, 89, 92, 105, 138
contact time, 75
contaminants, 36, 74, 96
contamination, 2, 6, 97
continuity, 46
control, 40, 44, 64, 67, 68, 69
convection, 150
convergence, 66, 67
conversion, 10, 12, 43, 69, 80, 98, 105, 108, 112, 139, 148, 149
conviction, 17
cooking, 20, 109, 111
cooling, 3, 4, 86, 109, 137, 138, 139, 144, 145, 148, 151, 152, 153, 154, 156, 159, 160, 161, 163, 166
Copenhagen, 34
copper, 95
coral reefs, 6
corn, 106
corporations, 4, 101
correlations, 84
corrosion, 163
corruption, 3
cosmetics, 103
cost effectiveness, 25
costs, 13, 160
credit, 16, 17
crime, 17
criminal acts, 17
criticism, 34
crops, 86, 106
crude oil, 20, 87, 92, 101, 107, 111, 112
crying, 84
crystal structure, 97
Cuba, 16

culture, 2, 3, 11, 35
curriculum, 33
cycles, 112, 137, 138, 139, 140, 147, 163

D

Darcy's law, 42, 46, 50, 53, 58, 64
death, 17, 22, 93
death rate, 93
decay, 127
decomposition, 30, 119, 120, 124, 127, 134, 135
decoupling, 75
defense, 17
deficiency, 1
definition, 7, 8, 17, 18, 19, 47, 145, 148, 150, 162
deforestation, 89, 90
degradation, 6, 30, 95, 107, 155
demand, 15, 99, 138
density, 42, 45, 46, 70, 141
Department of Energy, 115
depression, 20
derivatives, 45, 107
dermatitis, 118
desalination, 111
destruction, 1, 138, 159
detection, 37
developed countries, 100
developing countries, 8, 101
deviation, 128, 135
diabetes, 1, 3, 20, 103
dibenzofurans, 6
dibenzo-p-dioxins, 6
diet(s), 8, 20, 103
differential equations, 2, 3, 39, 40, 71, 119, 121, 134
diffusivities, 74
diffusivity, 91
dimensionality, 95
dioxin(s), 6, 38, 20, 107
disaster, 1, 16, 96
discharges, 31
discipline, 3
discretization, 39, 40, 42, 43, 44, 48, 49, 51, 62, 66, 67
disorder, 11
dispersion, 70
displacement, 70
distilled water, 74
distribution, 15, 16, 34, 38, 42, 50, 76, 87, 92, 95, 107
divergence, 18
diversity, vii
doctors, 17
dreaming, 84

drugs, 16
drying, 95
durability, 20, 21, 30, 102
duration, 17

E

ears, 160
earth, 6, 95, 97, 113, 138, 152, 159
eastern cultures, 95
ecology, 5, 14, 25, 34, 35, 36
economic activity, 13
economic development, 2, 6, 8, 87
economic growth, 12
economic indicator, 105
economic status, 12
economic sustainability, 12
economic theory, 11, 12, 13, 14
economics, 11, 14, 15, 34, 105, 157, 160
ecosystem, 6, 13, 20, 22, 25, 37, 159
education, 12, 16, 33, 35, 36, 38, 116
effluents, 73
Einstein, Albert, 2, 4, 84, 95, 137, 138, 139, 140, 141, 142, 144, 153, 154, 160, 163, 164, 165
elderly, 93
electric(al) power, 92, 98, 139
electric power transmission, 92
electricity, 4, 14, 89, 95, 106, 107, 108, 109, 112, 138, 139, 144, 145, 148, 149, 151, 156, 157, 160, 163
electricity system, 108
electrolysis, 105
electrolyte(s), 82
electromagnetic fields, 159
embargo, 17
emergency management, 17
emergency response, 16
emission, 30, 84, 89, 91, 92, 100, 101, 105, 107, 108, 109, 111, 113
energy, 2, 14, 22, 25, 27, 31, 34, 35, 36, 37, 83, 84, 85, 92, 95, 97, 98, 99, 100, 101, 103, 104, 105, 106, 107, 108, 109, 111, 112, 113, 117, 137, 138, 139, 144, 145, 149, 150, 151, 152, 153, 154, 155, 156, 157, 160, 162, 163, 165
energy consumption, 138
energy efficiency, 98, 100, 138, 157
Energy Information Administration, 114, 115
energy supply, 103
energy transfer, 153, 154, 156
Enhanced Oil Recovery, 117
entropy, 11

environment, 4, 5, 7, 8, 19, 25, 27, 30, 83, 85, 86, 87, 97, 98, 103, 107, 111, 113, 157, 158, 159, 160, 163
environmental chemicals, 34
environmental context, 8
environmental factors, 23
environmental impact, 19, 84, 98, 100, 105, 106, 108, 113
environmental sustainability, 35, 38
EPA, 89, 103, 105, 114
epidemic, 3, 20
equilibrium, 12
equipment, 15, 113, 139, 163
equity, 25
ethanol, 105, 106
ethers, 38
ethical questions, 22
ethylene (glycol), 37, 96
Euler, 54
Euro, 164
Eurocentric, 2
Europe, 86, 87
European, 28, 34, 36, 86, 87, 100, 138
European Union, 100
evacuation, 16
evaporation, 96
evidence, 16, 74, 83, 84, 87, 88, 101
evolution, 15
exclusion, 54, 148
execution, 16
exhaust heat, 98, 99
exporter, 16
exports, 13
exposure, 34, 159
external environment, 97
extinction, 6
extraction, 74, 152, 154, 157
extraction process, 152, 154

F

failure, 11, 101
faith, 85
famine, 86
farmers, 86
farms, 103
fats, 20, 99, 104
fears, 100
feedback, 19
feedstock, 92, 103, 105
fertilizer, 1, 20, 98, 99, 109, 111
fertilizers, 20
fever, 20

fiber(s), 20, 30, 34, 94
fibrils, 82
filtration, 82
financial support, 135
fire retardants, 37
fires, 91
First World, 87
fish, 2, 5, 13, 74, 75, 76, 77, 78, 80, 81, 99
Fish scale, 73
fisheries, 6, 13, 86
fishing, 13, 86, 91
flame, 34
flame retardants, 34
flexibility, vii
float, 141
flooding, 70
flue gas, 99, 109, 111
fluid, 40, 42, 43, 44, 45, 46, 50, 51, 53, 59, 60, 64, 67, 68, 70, 95, 109, 121, 138, 140, 144, 153, 154, 156, 159, 160
fluorine, 92
food, 3, 6, 14, 17, 22, 82, 86, 95, 112, 160
food production, 22
food products, 95
forest fire(s), 91
formaldehyde, 20, 96, 99, 105
fossil fuel(s), 20, 85, 88, 89, 91, 92, 94, 96, 97, 99, 101, 103, 104, 105, 108, 109, 111, 112, 138, 152, 153, 157, 163
free energy, 108
freedom, 1, 16
freezing, 95
Freud, 11
friction, 150
fruits, 17
fuel, 20, 84, 85, 87, 88, 89, 91, 92, 94, 96, 98, 99, 101, 103, 104, 105, 107, 108, 109, 111, 112, 113, 115, 138, 150, 152, 153, 156, 158, 159, 163
fulfillment, 100
funding, 8, 163

G

Gadus morhua, 74
gases, 89, 92, 93, 96, 100, 101, 102, 103, 111, 158
gasoline, 20, 105
General Motors, 138
generation, 89, 103, 107, 108, 113, 151
Georgia, 34, 164
glass, 155
GlaxoSmithKline, 7
global climate change, 102, 159
global forces, 90

Global Reporting Initiative, 7, 9, 35
global warming, 3, 83, 84, 85, 86, 87, 88, 89, 90, 91, 92, 93, 94, 96, 97, 100, 101, 105, 106, 111, 112, 113, 114, 117, 159
glucose, 22
glutamic acid, 80
glycol, 37, 95
glycolysis, 37
GNP, 28
goals, 9
goods and services, 87
government, 8, 9, 13, 16, 17, 28, 84, 90, 100
grants, 16
graph, 84, 161
gravity, 121
Great Britain, 86
green energy, 85
greenhouse gas, 83, 84, 88, 89, 92, 96, 100, 101, 102, 103, 107, 111, 113
Greenland, 86
grids, 43, 62, 63
gross domestic product (GDP), 2, 11, 12, 13, 14
ground water, 95
groups, 9, 74, 78, 80, 81, 87
growth, 7, 8, 11, 12, 13, 90
growth rate, 90
guidelines, 9, 10, 14, 83, 85, 113
Gulf Coast, 16
Gulf of Mexico, 16

H

harm, 94, 95
harmony, 95
Harvard, 15
harvesting, 89
Hawaii, 165
hazards, 6
health, 3, 11, 16, 37, 38, 94, 98, 99, 103, 107, 159
health effects, 37
health problems, 103
heart, 115, 116, 151
heat, 8, 91, 95, 96, 97, 98, 99, 108, 109, 112, 113, 128, 129, 137, 138, 139, 141, 144, 145, 146, 147, 148, 149, 150, 151, 152, 153, 154, 155, 156, 158, 159, 160
heat loss, 99, 151, 155, 156, 159
heat pumps, 160
heat release, 96
heat storage, 95
heat transfer, 96, 108, 144, 145, 150, 151, 154, 156
heating, 4, 96, 98, 99, 105, 108, 109, 111, 113, 139, 148, 149, 151, 152, 153, 156, 160

heavy metal(s), 2, 73, 74, 82, 95, 111, 112
hemisphere, 86
herbicide, 97
heterogeneity, vii, 74, 81
hexachlorobenzene, 6, 38
higher quality, 98
homogeneity, vii
human activity, 88
human brain, 3
human capital, 17
human condition, 1
human exposure, 34, 159
human milk, 38
humanity, 1, 6, 8, 17
hurricane(s), 16
hydrocarbon(s), 30, 39, 40, 87, 94
hydrogen, 74, 75, 80
hydrolysis, 105
hydrophobic, 78
hydroxide(s), 98, 104, 105, 109
hydroxyl, 74
hypothesis, 94

I

id, 117
Idaho, 73
identification, 163
impact assessment, 19
implementation, 28, 83, 113
imports, 17
imprisonment, 17
incidence, 138
income, 3, 8
independence, 139
independent variable, 119, 135
India, 20, 36, 87, 101, 117
indication, 148, 159
indicators, 6, 9, 10, 27, 28, 29, 35, 37, 38, 95, 105, 112
indirect effect, 159
Indonesia, 38
industrial application, 99, 109
industrial revolution, 17
industrial wastes, 97
industrialisation, 87
industrialized countries, 101
industry, 2, 10, 13, 14, 39, 40, 86, 89, 90, 92
inefficiency, 31
infinite, 1, 21, 22, 23, 95, 102, 105, 112, 119, 161, 162
Information Age, 1, 116
infrastructure, 87

inhibitors, 105
injustice, 1
innovation, 33, 35, 116, 135, 163, 164
insight, 4
instability, 119, 126, 134, 135
institutions, 6, 7
insulation, 37, 156
insulin, 3
integration, 15, 32
intensity, 80
intentions, 4, 16
interaction, 27, 85, 95, 96
interference, 144, 160
Intergovernmental Panel on Climate Change, 88, 100
internet, 84
interpretation, 10, 66, 67, 80
interrelationships, 15
interval, 42, 55, 120, 125, 130
intervention, 1, 85, 95, 97, 107, 111
intoxication, 87
investment, 12, 13, 16
investors, 101
ion adsorption, 74
ions, 73, 74, 76
Ireland, 82, 117
iron, 89
irradiation, 139
isomers, 6
isothermal, 146, 150
isotope(s), 83, 84, 91, 94, 108, 111, 113, 115
Italy, 114, 116

J

Japan, 2, 34, 36, 82, 100
Jordan, 35, 36, 116, 117, 165
Jung, 82
justification, 3, 39, 42, 66, 67, 87, 89, 101

K

Katrina, 16
Keynes, 2, 11
kidney, 104
killing, 13
kinetics, 35, 38
King, 64, 70
Kyoto Protocol, 83, 100, 101, 113, 117

L

labo(u)r, 15, 16

lakes, 95
land, 17, 88
Latin America, 87
laws, 18
lead, 12, 13, 54, 74, 76, 112, 161
leakage, 97
learning, 3
legislation, 138
life cycle, 10, 30, 31, 37, 105, 107
lifestyle, 1, 2, 3
limitation, 101, 105
linear function, 120
linear model, 102
literacy, 16
literature, 10, 66, 144
location, 139
London, 37, 38, 70
long-term impact, 85
Los Angeles, 71
love, 3
lung cancer, 20

M

macroeconomics, 12
magnesium, 92, 99
magnetic field, 159
magnetic resonance spectroscopy, 82
management, 5, 7, 17, 33, 35, 100, 109
manipulation, 87
manufacturer, 31
manufacturing, 36, 92, 99, 100, 107, 113
market, 74, 87
market share, 87
marketing, 13, 31, 32
markets, 15, 86
Marx, 15, 37
mathematics, 3, 39, 40, 64, 66, 162
matrix, 3, 10
meanings, 5
measurement, 8, 34, 150
measures, 16, 92, 100, 150, 153
mechanical energy, 150
media, 8, 40, 71, 82
membranes, 82
memory, 85, 95, 97
men, 20
messages, 114
metabolism, 94
metabolites, 6
metal ions, 73, 74
metals, 2, 82, 111, 112
metaphor, 15

methane, 93, 100, 104, 111, 112
methanol, 37, 104, 105
Mexico, 16
microbial, 90, 102, 114
microwave, 30
Middle East, 164
milk, 6, 36, 37, 38, 103
Millennium, 6, 8, 37
minerals, 20, 87, 98, 99
mining, 89, 108
missions, 83, 84, 88, 91, 92, 111, 113, 157
mixing, 95
modeling, 2, 36, 42, 70
models, 2, 7, 10, 15, 22, 33, 64, 83, 84, 85, 97, 100, 101, 102, 110, 113, 161
modern society, 20, 21, 103
modus operandi, 2
moisture, 96
molecular structure, 73, 80
molecules, 80
momentum, 95
money, 11, 12, 13, 15, 17, 33
monopoly, 87
morality, 33
Moscow, 73
motion, 109, 121
movement, 42, 95
multidimensional, 70
multiphase flow, 71
multiplication, 148
multiplicity, vii
myopia, 1

N

nanoparticles, 97
NASA, 92, 117
nation, 3, 13
natural environment, 4, 5, 27
natural food, 6, 22, 112
natural gas, 14, 88, 104, 111, 157
natural polymers, 82
natural resource management, 11, 12
natural resources, 25, 27
natural science, 2
Nepal, 37, 116
network, 75, 151
New Orleans, 16
New York, 35, 36, 38, 70, 82, 91, 116, 118, 135, 165, 166
New York Times, 91, 118
New Zealand, 34
Newton, 2

Newtonian, 3
nitric acid, 96
nitrogen, 73, 75, 80, 111
nitrous oxide, 93
non-linear, vii, 3, 18, 120
North America, 2, 17, 86, 87, 120
nuclear energy, 98, 100
nuclear magnetic resonance, 82
nuclear power plant, 112
nutrients, 20, 22
nutrition, 20

O

obesity, 1, 3, 20
obligation, 8
observations, 91
oceans, 84
offshore oil, 5, 28, 29, 30, 31, 35, 36
oil, 2, 5, 20, 28, 29, 30, 31, 35, 36, 39, 40, 64, 87, 88, 89, 92, 94, 98, 99, 101, 104, 105, 107, 108, 109, 111, 112, 113, 156, 157, 158, 160, 163
oil production, 158
oil refining, 89, 101, 112, 158
oligopolies, 87
one dimension, 128
opacity, 3
operator, 121, 132
organic compounds, 30
organism, 22
organization, 15
Organization for Economic Cooperation and Development, (OECD), 28, 37
organizations, 28, 84
osmosis, 82
osteoporosis, 103
overproduction, 15
ownership, 8, 15, 25
oxidants, 20
oxidation, 30, 37, 107
oxide(s), 93, 108, 111
oxygen, 75, 78, 95
ozone, 4, 95, 102, 115, 138, 159, 160, 164

P

pain, 3, 20
parabolic, 153, 154, 155, 156
parameter, 28, 74, 121, 161
Paris, 34, 37, 117
Parliament, 86

partial differential equations (PDE's), 2, 3, 39, 40, 71, 119, 121, 134
particles, 150
particulate matter, 99
pathways, 4, 15, 18, 30, 93, 97, 98
per capita income, 8
perception, 18, 161
performance, 7, 20, 35, 37, 39, 40, 137, 138, 141, 145
permeability, 43, 50, 67
personal, 16, 103
pesticide, 20, 97
pesticides, 1, 94, 97
petrochemical, 87
Petroleum, 36, 39, 70, 87, 114
petroleum products, 111
pH, 74
pharmaceutical, 91
phenol, 82
phosphate, 73, 75, 80, 81
phosphorus, 73, 75, 80, 81, 82
photosynthesis, 89, 91, 92, 94, 98, 99, 109, 113
photovoltaic, 112, 113, 138
physical environment, 85
physical properties, 74
physics, 39, 40, 135, 162
piracy, 84
plague, 36
planning, 17, 21
plants, 22, 89, 91, 92, 94, 99, 105, 111, 112, 150, 156
plastics, 20, 37, 87, 92
platinum, 37
pleasure, 3
polarization, 75, 82
politics, 3
pollutants, 6, 105
pollution, 91, 111, 157, 158, 159, 160, 163
polymers, 82
polynomials, 120, 126
polyurethane, 30, 37
polyurethane foam, 37
poor, 15, 155
population, 11, 13, 86, 109
porosity, 43, 45, 69, 74
porous media, 40, 71, 82
portability, 109
potassium, 98, 99, 104, 105, 109
potato, 103
poverty, 15
power, 75, 91, 92, 98, 107, 108, 112, 119, 120, 126, 139, 145, 149, 150, 151, 156, 158, 159, 166
power generation, 108
power plants, 91, 150, 156

precipitation, 88
prejudice, 17
preservatives, 1
pressure, 8, 16, 40, 42, 43, 48, 50, 53, 58, 60, 61, 63, 67, 68, 69, 70, 86, 95, 109, 137, 138, 139, 140, 142, 158, 163
prevention, 37
probe, 75
producers, 16
product life cycle, 107
production, 13, 14, 15, 16, 22, 30, 35, 39, 40, 51, 61, 84, 85, 87, 89, 91, 95, 98, 101, 103, 104, 105, 106, 107, 108, 111, 112, 113, 138, 156, 158
profit, 21
profitability, 87
programming, 39, 40
propaganda, 87
proposition, 84
prosperity, 3
prostate, 37
prostate cancer, 37
protein(s), 73, 75, 78, 79, 80
protocol, 101
prototype, 144
Prozac, 20
Psychoanalysis, 11
psychology, 3
public opinion, 84, 87
public sector, 31
pumps, 144, 160
pure water, 96, 97
purification, 73, 95

Q

questioning, 93

R

race, 2
radiation, 90, 91, 96, 100, 114, 118, 138, 150, 155, 159, 160
rain, 97, 157
range, 6, 12, 16, 19, 78, 80, 132, 135, 151, 153
rape, 15
raw materials, 15, 87
Reagan Administration, 8
reality, 20, 22, 66, 102, 150, 161
recall, 10
recovery, 39, 98
recycling, 36, 38
reduction, 18, 37, 100, 101

refining, 87, 89, 91, 94, 101, 103, 107, 108, 111, 112, 113, 158
reflection, 21, 34, 62, 63, 97, 159
refrigeration, 4, 109, 137, 138, 139, 140, 141, 142, 144, 145, 147, 148, 149, 151, 152, 156, 160, 162, 163
regeneration, 111
regional, 102
regulatory requirements, 25
rejection, 82
relationship, 64
renewable energy, 31, 152
Renewable Energy Technology, 37
reputation, 16
residues, 78, 79, 80
resistance, 151
resolution, 70, 82
resource management, 11, 12
resources, 7, 8, 13, 25, 27, 31, 35, 38, 115
respiratory, 13
retail, 16
retardation, 74
returns, 16
revenue, 13
Rio de Janeiro, 5, 8
risk, 38
rivers, 2, 95
Rome, 114
rural population, 86

S

salt, 105
sample, 78, 79, 80, 81
saturated fat, 20
saturation, 140
savings, 16
sawdust, 111
scandal, 11
school, 15, 16
science, 1, 2, 3, 4, 33, 35, 84, 85, 87, 97, 162
sea level, 92, 100
seafood, 2
sea-level rise, 92
Second World, 8
security, 8
sediments, 30
selecting, 33, 35, 112
self-destruction, 1
self-interest, 161
SEM, 73, 74, 75, 76, 80, 81
sensitivity, 91

series, 34, 40, 80, 83, 84, 101, 107, 111, 113, 116, 119, 120, 121, 122, 123, 124, 125, 126, 127, 128, 129, 130, 131, 132, 133, 134, 135
sewage, 17, 95, 111
shape, 97, 155
shares, 87
Shell, 7
shock, 118
shoot, 117
sign, 13, 23
signals, 78
silica, 98, 99
silicon, 112
silk, 78, 82
similarity, 80
simulation, 2, 23, 36, 39, 42, 54, 64, 65, 66, 70, 71
sine, 132, 134
sites, 2, 74, 80
skin, 20, 103
skin cancer, 20
smoke, 13
social benefits, 105
social context, 7
social problems, 20
social responsibility, 16
social theory, 10
society, 2, 3, 8, 15, 16, 17, 20, 21, 34, 82, 87, 88, 103
sodium, 74, 98, 99, 104, 105, 109
sodium hydroxide, 98, 104, 105
software, 134
soil, 20, 22, 160
solar, 90, 96, 98, 100, 105, 108, 109, 112, 113, 137, 138, 139, 144, 153, 154, 155, 156, 157, 160, 162, 163
solar cell(s), 113
solar collection, 160
solar energy, 98, 100, 105, 108, 113, 137, 138, 139, 144, 153, 154, 155, 156, 160, 162, 163
solar system, 137, 154
solid phase, 74
solid state, 82
solitons, 120
solubility, 78
solvent, 95
sorption, 74
sorting, 23
Soviet Union, 16
Spain, 35
species, 6, 15, 74, 80
spectroscopy, 73, 82
spectrum, 75, 78, 79, 80, 81
speech, 35, 116
speed, 134

sperm, 20
stability, 66, 67, 71, 120, 124, 134
stack gas, 150, 158
stages, 23, 32, 98, 105, 158, 159
standards, 84, 101, 103
steel, 89, 95
Stochastic, 135
stock, 13, 104
storage, 95, 108, 144
stoves, 98, 100, 109, 111
strategies, 8, 39
strength, vii, 10
structural protein, 78
substitutes, 4
substitution, 48, 105
substrates, 74
Sudan, 97, 118
sugar, 1, 20, 75
suicidal behavior, 20
sulfur, 92, 93, 108
sulfuric acid, 104, 105
summer, 8, 16, 86, 156
superiority, 15
supply, 15, 30, 31, 32, 33, 36, 90, 103, 144, 156
supply chain, 30, 31, 32, 33, 36
surface area, 74, 111, 155, 156
surfactant, 98, 158
surplus, 16
sustainability, vii, 2, 5, 6, 7, 8, 9, 10, 12, 14, 19, 21, 23, 25, 27, 28, 29, 30, 31, 33, 34, 35, 36, 37, 38, 83, 85, 102, 105, 106, 113, 161, 162
sustainable development, 1, 5, 6, 8, 9, 10, 16, 17, 25, 28, 33, 35, 37, 38, 100
sweets, 20
swelling, 80
switching, 105
Switzerland, 164
sympathy, 86
symptom(s), 1, 2, 20, 84
synthesis, 37
synthetic fiber, 94
systems, 10, 14, 15, 16, 17, 28, 37, 66, 85, 95, 98, 102, 107, 112, 113, 137, 138, 139, 144, 150, 152, 160, 161

T

Taiwan, 115, 116
tanks, 95
targets, 100, 101
tariffs, 86
Taylor series, 40, 126
teaching, 64

technological developments, 5, 35
technology, vii, 1, 2, 5, 6, 7, 11, 14, 18, 20, 21, 22, 23, 24, 25, 26, 27, 29, 30, 31, 32, 33, 35, 83, 85, 96, 98, 100, 102, 103, 105, 106, 107, 108, 109, 111, 112, 113, 137, 138, 161, 162
temperature, 42, 84, 91, 97, 100, 101, 102, 107, 137, 139, 140, 142, 143, 144, 146, 147, 150, 152, 153, 154, 155, 156
temperature gradient, 144
temporal, 6, 7, 33, 36
terminals, 80
Texas, 36, 70
textiles, 38
theory, 10, 11, 12, 13, 14, 83, 84, 94, 96
thermal energy, 144, 150, 163
thermodynamic(s), 138, 141, 142, 144, 146, 150
thinking, 2, 10, 39, 64, 66, 162
third boundary condition, 58
threat, 19, 33
time frame, 21
time variables, 119
tobacco, 1, 13
toluene, 105
toxic contamination, 6
toxic gases, 158
toxic products, 93
toxicity, 38, 103, 158
toxins, 2
trade, 17, 138
trading, 100, 101
transesterification, 99, 104
transformation, 35
transition, 2
transmission, 15, 16, 92, 107, 108, 151, 153, 154, 159, 166
transport, 82, 85, 92, 94, 96
transportation, 30, 31, 89, 92, 95, 97
trend, 6, 12, 33, 36, 90, 101, 104, 107, 126
tribes, 86
tumors, 94
turbulence, 121, 127
Turkey, 35, 36

U

UK, 100, 114, 115
ultimate analysis, 73, 74, 75, 80
UN, 5, 8, 10, 28
UN Conference on Environment and Development (UNCED), 5, 8
UNESCO, 34
United Arab Emirates (U.A.E.), 14, 35, 39
United Nations, 7, 8, 9, 28, 37, 38, 117

United States, 8, 9, 16
univariate, 84
universe, 162
universities, 35
uranium, 108, 118
urbanization, 89
users, 2
UV, 95

V

vacuum, 91
validation, 39, 40
validity, 53
values, 40, 43, 49, 125, 134, 135, 161
vapor, 96, 97, 137, 138, 139, 141, 143, 145, 146, 147, 148, 149, 151, 152, 159, 160, 163
variable(s), 15, 51, 52, 84, 95, 119, 122, 123, 129, 134, 135
variation, 51
vegetable oil, 94, 99, 104, 105, 107, 108, 111, 156, 160, 162
vegetables, 17, 112
velocity, 43, 45, 53, 69
Venezuela, 70
vertical integration, 15
vessels, 13
victims, 17
Vioxx, 20
viscosity, 43, 50, 70, 121
vision, 21, 27
vitamin C, 93, 114, 115
vitamins, 94, 103

W

wages, 15, 86
war, 96
Washington, 37, 114, 116, 164
waste management, 100
waste treatment, 107
water vapor, 96, 97, 143
wealth, 15
weapons, 13
web, 29, 30
well-being, 6
wells, 42
wholesale, 16
winning, 86
winter, 8, 144
wood, 1, 20, 94, 98, 99, 100, 105, 108, 109, 110, 111, 113, 160

wool, 30
workers, 16, 80, 94
World Bank, 28
World Health Organization (WHO), 38
World War, 8, 87
writing, 46, 51, 65

X

xylene, 105

Y

yield, 13, 20, 42

Z

zero-waste, 83, 110, 111, 113
Zimbabwe, 114
zooplankton, 22